FUNDAMENTALS
OF
CLINICAL TRIALS

SECOND EDITION

Fundamentals
of
Clinical Trials

Second Edition

Lawrence M. Friedman, MD
Curt D. Furberg, MD
David L. DeMets, PhD

PSG PUBLISHING COMPANY, INC.
LITTLETON, MASSACHUSETTS

Library of Congress Cataloging in Publication Data

Friedman, Lawrence M.   1942–
    Fundamentals of clinical trials.

    Includes bibliographies and index.
    1. Clinical trials.   I. Furberg, Curt., 1936–
II. DeMets, David L., 1944–   . III. Title.
[DNLM: 1. Clinical Trials.   2. Research Design.
W 20.5 F911f]
R853.C55F75   1985   615.5′0724   85-3559
ISBN 0-88416-499-3

Published by:
PSG PUBLISHING COMPANY, INC.
545 Great Road
Littleton, Massachusetts 01460

First Edition 1981
Second Edition 1985

This publication results in part from activities undertaken by the authors and other staff at the National Heart, Lung, and Blood Institute.

International Standard Book Number: 0-88416-499-3

Library of Congress Catalog Card Number: 85-3559

Last digit is the print number: 9 8 7 6 5 4 3

ABOUT THE AUTHORS

Lawrence M. Friedman, M.D., is Acting Chief of the Clinical Trials Branch, National Heart, Lung, and Blood Institute, National Institutes of Health. He has had major involvement in several large clinical trials and has been a consultant on a number of others. He has a research interest in clinical trials methodology.

Curt D. Furberg, M.D., is Professor of Medicine, Director of the Center for Prevention Research and Biometry, Bowman Gray School of Medicine and former Associate Director, Clinical Applications and Prevention Program, National Heart, Lung, and Blood Institute, National Institutes of Health. He has played major scientific and administrative roles in several large-scale clinical trials and has had consultative activities on others. Dr. Furberg's research activities include the area of clinical trials methodology and cardiovascular epidemiology.

David L. DeMets, Ph.D. in Biostatistics, is Professor of Statistics and Human Oncology, Director of Biostatistics, University of Wisconsin, and former Chief of the Mathematical and Applied Statistics Branch, National Heart, Lung, and Blood Institute, National Institutes of Health. He has had major roles in the design, monitoring and analysis of data from a number of clinical trials, especially in the pulmonary disease area. Since 1982, he has been involved in clinical trials of cancer therapy. Recent research interests involve monitoring and early stopping rules in clinical trials.

CONTENTS

PREFACE

The clinical trial is "the most definitive tool for evaluation of the applicability of clinical research." It represents "a key research activity with the potential to improve the quality of health care and control costs through careful comparison of alternative treatments."[1] In 1979, by itself the National Institutes of Health in the United States supported almost 1000 clinical trials at a cost of greater than $136 million.

While many reported clinical trials are of high quality, a careful reviewer of the medical literature will notice that a large number have deficiencies in design, conduct, analysis, presentation, or interpretation of results. Although improvements have occurred in clinical studies,[2] there is still considerable doubt that major progress has taken place over the past thirty years.[3-5] Certainly, many studies could have been upgraded if the authors had had a better understanding of the fundamentals.

Since the publication of the first edition of this book, several other texts on clinical trials have appeared.[6-11] Few of them, however, discuss general as well as specific issues involved in the development, management, and analysis of clinical trials. The purpose of this second edition is to update areas in which considerable progress has recently been made and to add chapters on topics which have not received adequate attention in the past.

In this book we hope to assist investigators in improving the quality of clinical trials by discussing fundamental concepts with examples from our experience and the literature. It is intended primarily for investigators with some clinical trial experience as well as for those who plan to conduct a trial for the first time. It may also be used in the teaching of clinical trial methodology and to assist members of the scientific and medical community who wish to evaluate and interpret published reports of trials.

The first half of the book concerns design and development phases of a trial. These chapters are relevant for all investigators. Chapter 9 reviews recruitment techniques and may be of special interest to investigators not having ready access to subjects. Methods for collecting high quality data as well as some common problems in data collection are included in Chapter 10. Chapters 11 and 12 focus on the important areas of assessment of adverse effects and quality of life. Measures to enhance and monitor subject compliance are presented in Chapter 13. Chapter 14 reviews techniques of survival analysis. Chapter 15 covers data monitoring. Which subjects should be analyzed? The authors develop this question posed in Chapter 16 by discussing reasons for not withdrawing subjects from analysis. Topics such as subgroup analysis are also addressed. Chapter 17 deals with phasing out clinical trials, and Chapter 18 with reporting and interpretation of results. Finally, in Chapter 19 the text presents information about multicenter studies, which have features requiring special attention. Several points covered in the final chapter may also be of value to investigators conducting single center studies.

This book is a collaborative effort and is based on knowledge gained during more than a decade in developing, conducting, overseeing, and analyzing data from a number of clinical trials. This experience is chiefly, but not exclusively, in large trials of heart and lung diseases and cancer. As a consequence, many of the examples cited are based on work done in these fields. However, the principles are applicable to clinical trials in general.

The views expressed in this book are those of the authors and do not necessarily represent the views of the National Heart, Lung, and Blood Institute.

REFERENCES

1. NIH Inventory of Clinical Trials: Fiscal Year 1979. Volume I. National Institutes of Health, Division of Research Grants, Research Analysis and Evaluation Branch, Bethesda, MD.

2. Chalmers, T.C., and Schroeder, B. Letter to the editor. *N Engl J Med.* 301:1293, 1979.

3. Ross, O.B., Jr. Use of control in medical research. *JAMA* 145:72–75, 1951.

4. Juhl, E., Christensen, E., and Tygstrup, N. The epidemiology of the gastrointestinal randomized clinical trial. *N Engl J Med.* 296:20–22, 1977.

5. Fletcher, R.H., and Fletcher, S.W. Clinical research in general medical journals: a 30-year perspective. *N Engl J Med.* 301:180–183, 1979.

6. Tygstrup, N., Lachin, J.M., and Juhl, E. (Eds): *The Randomized Clinical Trial and Therapeutic Decisions.* New York: Marcel Dekker, 1982.

7. Miké, V. and Stanley, K.E. (Eds): *Statistics in Medical Research: Methods and Issues, With Applications in Cancer Research.* New York: John Wiley and Sons, 1982.

8. Ingelfinger, J.A., Mosteller, F., Thibodeau, L.A., et al. *Biostatistics in Clinical Medicine.* New York: Macmillan, 1983.

9. Bulpitt, C.J. *Randomised Controlled Clinical Trials.* The Hague: Martinus Nijhoff, 1983.

10. Pocock, S.J. *Clinical Trials: A Practical Approach.* New York: John Wiley and Sons, 1983.

11. Shapiro, S.H., and Louis, T.A. (Eds): *Clinical Trials: Issues and Approaches.* New York: Marcel Dekker, 1983.

ACKNOWLEDGMENTS

Most of the ideas and concepts discussed in this book represent what we have learned during our years at the National Heart, Lung, and Blood Institute. We are indebted to many colleagues, particularly Dr. Max Halperin and Dr. William T. Friedewald, with whom we have had numerous hours of discussion concerning theoretical and operational aspects of the design, conduct and analysis of clinical trials.

Many contributed to the preparation of the first edition of this book. We are especially grateful for the advice and assistance of Dr. Byron W. Brown, Dr. Max Halperin, Dr. Darwin R. Labarthe, Dr. Graham May, and Dr. Hans Wedel. We would also like to thank Dr. Thomas C. Chalmers, Dr. William T. Friedewald, Dr. Margaret E. Mattson, and Dr. Daniel G. Seigel for their review. We appreciate the many constructive comments of Dr. Labarthe and Dr. May in preparing this second edition.

CHAPTER 1

Introduction to Clinical Trials

The evolution of the clinical trial dates back to the 18th century.[1,2] Lind, in his classical study on board the *Salisbury*, evaluated six treatments for scurvy in 12 patients. One of the two who was given oranges and lemons recovered quickly and was fit for duty after six days. The second was the best recovered of the others and was assigned the role of nurse to the remaining ten patients. Several other comparative studies were also conducted in the 18th and 19th centuries. The comparison groups comprised literature controls, other historical controls and concurrent controls.[2]

The concept of randomization was introduced by Fisher and applied in agricultural research in 1926.[3] The first clinical trial that used a form of random assignment of subjects to study groups was reported in 1931 by Amberson et al.[4] After careful matching of 24 patients with pulmonary tuberculosis into comparable groups of 12 each, a flip of a coin determined which group received sanocrysin, a gold compound commonly used at that time. The British Medical Research Council trial of streptomycin in patients with tuberculosis, reported in 1948, was the first to use random numbers in the allocation to experimental and control groups.[5]

The principle of blindness was also introduced in the trial by Amberson et al[4] mentioned above. The patients were not aware of whether they received intravenous injections of sanocrysin or distilled water. In a trial of cold vaccines in 1938, Diehl and coworkers[6] referred to the saline solution given to the subjects in the control group as a placebo.

It is only in the past three decades that the clinical trial has emerged as the preferred method in the evaluation of medical interventions. Techniques of implementation and special methods of analysis have been developed during this period. Many of the principles have their origins in work by Hill.[7-10]

The purpose of this chapter is to define clinical trials, review the need for them and discuss timing and ethics of clinical trials.

FUNDAMENTAL POINT

A properly planned and executed clinical trial is a powerful experimental technique for assessing the effectiveness of an intervention.

WHAT IS A CLINICAL TRIAL?

A clinical trial is defined as a *prospective study comparing the effect and value of intervention(s) against a control in human subjects.* Note that a clinical trial is *prospective*, rather than retrospective. Study subjects must be followed forward in time. They need not all be followed from an identical calendar date. (In fact, this will occur only rarely.) Each subject however, must be followed from a well-defined point, which becomes time zero or baseline for the study. This contrasts with a case-control study, a type of retrospective study in which subjects are selected on the basis of presence or absence of an event of interest. By definition, such a study is not a clinical trial. Subjects can also be identified from hospital records or other data sources and subsequent records can be assessed for evidence of events. This is not considered to be a clinical trial since the subjects are not directly observed from the moment of initiation of the study and at least some of the follow-up data are retrospective.

A clinical trial must employ one or more *intervention* techniques. These may be "prophylactic, diagnostic or therapeutic agents, devices, regimens, procedures, etc."[11] Intervention techniques should be applied to subjects in a standard fashion in an effort to change some aspect of the subjects. Follow-up of subjects over a period of time without active intervention may measure the natural history of a disease process, but it does not constitute a clinical trial. Without active intervention the study is observational because no experiment is being performed.

A clinical trial must contain a *control* group against which the intervention group is compared. At baseline, the control group must be sufficiently similar in relevant respects to the intervention group in order that differences in outcome may reasonably be attributed to the action of the intervention. Methods for obtaining an appropriate control group are discussed in Chapter 4. Most often a new intervention is compared with best current standard therapy. If no such standard exists, the subjects in the intervention group may be compared to subjects who are on no active intervention. "No active intervention" means that the subject may receive either a placebo or no intervention at all. Obviously, subjects in all groups may be on a variety of additional non-standard therapies and regimens, so called concomitant interventions, which may be either self-administered or prescribed by others (eg, private physicians).

In drug development, a phase I study is done to evaluate maximum tolerated dose and toxicity. A phase II study assesses drug activity, in addition to toxicity. Unless control groups are employed, these studies do not meet the definition of a clinical trial used in this book. This book addresses issues related to phase III studies, which are comparative.

For purposes of this book, only studies on *human beings* will be considered as clinical trials. Certainly, animals (or plants) may be studied using similar techniques. However, this book focuses on trials in people, and each clinical trial must therefore incorporate subject safety considerations into its basic design. Equally important is the need for, and responsibility of, the investigator to fully inform potential subjects about the trials.[12,13]

Unlike animal studies, in clinical trials the investigator cannot dictate what an individual should do. He can only strongly encourage subjects to avoid certain medications or procedures which might interfere with the trial. Since it may be impossible to have "pure" intervention and control groups, an investigator may not compare interventions, but only intervention strategies. Strategies refer to attempts at getting all subjects to comply to the best of their ability with their originally assigned intervention. When planning a trial, the investigator should recognize the difficulties inherent in studies with human subjects and attempt to estimate the magnitude of subjects' failure to comply strictly with the protocol.

As discussed in Chapters 5 and 6, *the ideal clinical trial is one that is randomized and double-blind.* Deviation from this standard has potential drawbacks which will be discussed in the relevant chapters. In some clinical trials compromise is unavoidable, but often deficiencies can be prevented by adhering to fundamental features of design and conduct.

WHY ARE CLINICAL TRIALS NEEDED?

A clinical trial is the most definitive method of determining whether an intervention has the postulated effect. Only seldom is a disease or condition so completely characterized that people fully understand its natural history and can say, from a knowledge of pertinent variables, what the subsequent course of a group of patients will be. Even more rarely can a clinician predict with certainty the outcome in individual patients. By outcome is meant not simply that an individual will die, but when, and under what circumstances; not simply that he will recover from a disease, but what complications of that disease he will suffer; not simply that some biological variable has changed, but to what extent the change has occurred. Given the uncertain knowledge about disease course and the usual large variations in biological measures, it is usually

4

impossible to say on the basis of uncontrolled clinical observation whether a new treatment has made a difference to outcome. A clinical trial offers the possibility of such judgment because there exists a control group—which, ideally, is comparable to the intervention group in every way except for the intervention being studied.

The consequences of not conducting appropriate clinical trials at the proper time can be serious or costly. An example is the continued uncertainty as to the efficacy and safety of digitalis in congestive heart failure. Intermittent positive pressure breathing became an established therapy for chronic obstructive pulmonary disease without good evidence of benefits. One recent trial suggested no major benefit from this very expensive procedure.[14] Similarly, high concentration of oxygen was used for therapy in premature infants until a clinical trial demonstrated its harm.[15]

A clinical trial can determine the incidence of adverse effects or complications of the intervention. Few interventions, if any, are entirely free of undesirable effects. However, drug toxicity might go unnoticed without the systematic follow-up measurements obtained in a clinical trial. Chapters 11 and 12 cover the assessment of this in more detail.

In the final evaluation, an investigator must compare the benefit of an intervention with its other, possibly unwanted effects in order to decide whether, and under what circumstances, its use should be recommended. Such assessments are not statistical in nature. They must rely on the judgment of the investigator.

There is no such thing as a perfect study. However, a well thought-out, well-designed, appropriately conducted and analyzed clinical trial is an effective tool. While even well designed clinical trials are not infallible, they can provide a sounder rationale for intervention than is obtainable by other methods of investigation. On the other hand, poorly designed and conducted trials can be misleading. Also, without supporting evidence, no single study ought to be definitive. When interpreting the results of a trial, consistency with data from laboratory, animal, epidemiological, and other clinical research must be considered.

PROBLEMS IN THE TIMING OF A TRIAL

Once drugs and procedures of questionable benefit have become part of general medical practice, performing an adequate clinical trial becomes difficult ethically and logistically. An example of this is the issue concerning use of anticoagulants after a myocardial infarction. Whether or not the chronic use of anticoagulants in people who have suffered a myocardial infarction is helpful or harmful is unclear. This controversy may never be settled. A properly performed clinical trial done before widespread use of anticoagulants might have spared lives and would certainly have saved considerable money and time.

Some people advocate instituting clinical trials as early as possible in the evaluation of new therapies.[16] The trials, however, must be feasible. Assessing feasibility takes into account several factors. Before conducting a trial, an investigator needs to have the necessary knowledge and tools. He must know something about the safety of the intervention. He needs to know what outcomes to assess and have the techniques to do so. Well run clinical trials of adequate magnitude are costly and should be done only when preliminary evidence of the efficacy of an intervention looks promising enough to warrant the effort and expense involved.

Another aspect of timing is consideration of the relative stability of the intervention. If active research will be likely to make the intended intervention outmoded in a short time, studying such an intervention may be inappropriate. This is particularly true in long-term clinical trials, or studies that take many months to develop. One of the criticisms of trials of surgical interventions has been that surgical methods are constantly being improved. Evaluating an operative technique of several years past, when a study was initiated, may not reflect the current status of surgery.[17,18]

These issues were raised in connection with the Veterans Administration study of coronary artery bypass surgery.[19] The trial showed that surgery was beneficial in subgroups of patients with left main coronary artery disease and three vessel disease, but not overall.[19-21] Critics of the trial argued that when the trial was started, the surgical techniques were still evolving. Therefore, surgical mortality in the study did not reflect what occurred in actual practice at the end of the long-term trial. In addition, there were wide differences in surgical mortality between the cooperating clinics,[22] which may have been related to the experience of the surgeons. Defenders of the study maintained that the surgical mortality in the Veterans Administration hospitals was not very different from the national experience at the time.[23] In the more recent Coronary Artery Surgery Study,[24] surgical mortality was lower than in the Veterans Administration trial, reflecting better technique. The control group mortality, however, was also lower.

Review articles show that surgical trials have been successfully undertaken.[25,26] While the best approach would be to postpone a trial until a procedure has reached a plateau and is unlikely to change greatly, such a postponement will probably mean waiting until the procedure has been widely accepted as efficacious, thus making it impossible to conduct the trial. However, as noted by Chalmers and Sacks,[27] allowing for improvements in operative techniques in a clinical trial is possible. As in all aspects of conducting a clinical trial, judgment must be used in determining the proper time to evaluate an intervention.

ETHICS OF CLINICAL TRIALS

People have debated the ethics of clinical trials for as long as they have been done. The arguments have changed over the years and perhaps become more sophisticated, but in general, they center around the issues of the physician's obligations to his patient versus societal good, informed consent, randomization, and the use of placebo.[28-39] Studies that require ongoing intervention or studies that continue to enroll subjects after trends in the data have appeared have raised some of the controversy.[29,40,41] The indicated references argue a number of these issues.

The authors of this text take the view that properly designed and conducted clinical trials are ethical. A well-designed trial can answer important public health questions without impairing the welfare of individuals. There may, at times, be conflicts between a physician's perception of what is good for his patient, and the needs of the trial. In such instances, the needs of the subject must predominate. Proper informed consent is essential. The requirements of the U.S. Department of Health and Human Services are reasonable ones.[12] A number of investigators have shown that simply adhering to legal requirements does not ensure informed consent.[42,43] In many clinical trial settings, though, true informed consent can be obtained.[44] The situations where subject enrollment must be done immediately or in highly stressful circumstances and where the prospective subjects are minors or not fully competent to understand the study are more complicated and may not have optimal solutions.

Randomization has generally been more of a problem for physicians and investigators than for subjects.[45] The objection to random assignment should only apply if the investigator believes that a preferred therapy exists. If that is the case, he should not participate in the trial. On the other hand, if he truly cannot say that one treatment is better than another, there should be no ethical problem with randomization. Such judgments regarding efficacy obviously vary among investigators. Similarly, the use of a placebo is acceptable if there is no known best therapy. Of course, all subjects must be told that there is a specified probability, eg, 50%, of their receiving placebo. The use of a placebo also does not imply that control group subjects will receive no treatment. In many trials, the objective is to see whether a new intervention plus standard care is better or worse than a placebo plus standard care.

The issue of how to handle accumulating data from an ongoing trial is a difficult one, and is discussed in Chapter 15. With advance understanding by both subjects and investigators that they will not be told interim results, and that there is a responsible data monitoring group, ethical concerns should be lessened, if not totally alleviated.

STUDY PROTOCOL

Every well-designed clinical trial requires a protocol. The study protocol can be viewed as a written agreement between the investigator, the subject, and the scientific community. The contents provide the background, specify the objectives, and describe the design and organization of the trial. Every detail explaining how the trial is carried out does not need to be included, provided that a comprehensive manual of procedures contains such information. The protocol serves as a document to assist communication among those working in the trial. It should also be made available to others upon request.

The protocol should be developed before the beginning of subject enrollment and should remain essentially unchanged except perhaps for minor updates. Careful thought and justification should go into any changes. Major revisions which alter the direction of the trial should be rare. An outline of a protocol is given below:

A. Background of the study
B. Objectives
    1. Primary question and response variable
    2. Secondary questions and response variables
    3. Subgroup hypotheses
C. Design of the study
    1. Study population
        a. Inclusion and exclusion criteria
        b. Sample size estimates
    2. Enrollment of subjects
        a. Informed consent
        b. Assessment of eligibility
        c. Baseline examination
        d. Intervention allocation
    3. Intervention
        a. Description and schedule
        b. Measures of compliance
    4. Follow-up visit description and schedule
    5. Ascertainment of response variables
        a. Training
        b. Data collection
        c. Data monitoring and quality control
        d. Data analysis
        e. Termination policy
D. Organization
    1. Participating investigators

2. Study administration
   a. Committees and subcommittees
   b. Policy and Data Monitoring Committee

Appendix
   Definitions of entrance criteria
   Definitions of response variables

REFERENCES

1. Bull, J.P. The historical development of clinical therapeutic trials. *J Chronic Dis.* 10:218–248, 1959.

2. Lilienfeld, A.M. Ceteris paribus: the evolution of the clinical trial. *Bull History Medicine* 56:1–18, 1982.

3. Box, J.F. R.A. Fisher and the design of experiments, 1922–1926. *Am Stat.* 34:1–7, 1980.

4. Amberson, J.B., Jr., McMahon, B.T., and Pinner, M. A clinical trial of sanocrysin in pulmonary tuberculosis. *Am Rev Tuberculosis* 24:401–435, 1931.

5. Medical Research Council. Streptomycin treatment of pulmonary tuberculosis. *Br Med J.* 2:769–782, 1948.

6. Diehl, H.S., Baker, A.B., and Cowan, D.W. Cold vaccines; an evaluation based on a controlled study. *JAMA* 111:1168–1173, 1938.

7. Hill, A.B. The clinical trial. *Br Med Bull.* 7:278–282, 1951.

8. Hill, A.B. The clinical trial. *N Engl J Med.* 247:113–119, 1952.

9. Hill, A.B. *Statistical Methods of Clinical and Preventive Medicine.* New York: Oxford University Press, 1962.

10. Doll, R. Clinical trials: retrospect and prospect. *Stat Med.* 1:337–344, 1982.

11. NIH Inventory of Clinical Trials: Fiscal Year 1979. Volume I. National Institutes of Health, Division of Research Grants, Research Analysis and Evaluation Branch, Bethesda, MD.

12. OPRR Reports. Code of Federal Regulations: (45 CFR 46) Protection of Human Subjects. National Institutes of Health, Department of Health and Human Services. Revised as of March 8, 1983.

13. National Commission for the Protection of Human Subjects of Biomedical and Behavioral Research. The Belmont Report: ethical principles and guidelines for the protection of human subjects of research. *Federal Register* 44:23192–23197, 1979.

14. The Intermittent Positive Pressure Breathing Trial Group, Intermittent positive pressure breathing therapy of chronic obstructive pulmonary disease—a clinical trial. *Ann Intern Med.* 99:612–620, 1983.

15. Silverman, W.A. The lesson of retrolental fibroplasia. *Sci Am.* 236:100–107, June 1977.

16. Chalmers, T.C. Randomization of the first patient. *Med Clin North Am.* 59:1035–1038, 1975.

17. Bonchek, L.I. Are randomized trials appropriate for evaluating new operations? *N Engl J Med.* 301:44–45, 1979.

18. Van der Linden, W. Pitfalls in randomized surgical trials. *Surgery* 87:258–262, 1980.

19. Murphy, M.L., Hultgren, H.N., Detre, K., et al. Treatment of chronic stable angina — a preliminary report of survival data of the randomized Veterans Administration cooperative study. *N Engl J Med.* 297:621–627, 1977.

20. Takaro, T., Hultgren, H.N., Lipton, M.J., et al. The VA Cooperative Study of Surgery for Coronary Arterial Occlusive Disease. II. Subgroup with significant left main lesions. *Circulation* 54 (Suppl III):III-107–III-117, 1976.

21. Detre, K., Peduzzi, P., Murphy, M., et al. Effect of bypass surgery on survival in patients in low- and high-risk subgroups delineated by the use of simple clinical variables. *Circulation* 63:1329–1338, 1981.

22. Proudfit, W.L. Criticisms of the VA randomized study of coronary bypass surgery. *Clin Res.* 26:236–240, 1978.

23. Chalmers, T.C., Smith, H., Jr., Ambroz, A., et al. In defense of the VA randomized control trial of coronary artery surgery. *Clin Res.* 26:230–235, 1978.

24. CASS Principal Investigators and Their Associates. Myocardial infarction and mortality in the Coronary Artery Surgery Study (CASS) randomized trial. *N Engl J Med.* 310:750–758, 1984.

25. Strachan, C.J.L., and Oates, G.D. Surgical trials. Edited by F.N. Johnson and S. Johnson. In *Clinical Trials.* Oxford: Blackwell Scientific, 1977.

26. Bunker, J.P., Hinkley, D., and McDermott, W.V. Surgical innovation and its evaluation. *Science* 200:937–941, 1978.

27. Chalmers, T.C., and Sacks, H. Letter to the editor. *N Engl J Med.* 301:1182, 1979.

28. Schafer, A. The ethics of the randomized clinical trial. *N Engl J Med.* 307:719–724, 1982.

29. Marquis, D. Leaving therapy to chance. *The Hastings Center Report* 13:40–47, August 1983.

30. Burkhardt, R., and Kienle, G. Controlled clinical trials and medical ethics. *Lancet* ii:1356–1359, 1978.

31. Rutstein, D.D. The ethical design of human experiments. Edited by P.A. Freund. In *Experimentation with Human Subjects.* New York: George Braziller, 1970.

32. Shaw, L.W., and Chalmers, T.C. Ethics in cooperative clinical trials. *Ann NY Acad Sci.* 169:487–495, 1970.

33. Vere, D.W. Ethics of clinical trials. Edited by C.S. Good. In *The Principles and Practice of Clinical Trials.* Edinburgh: Churchill Livingstone, 1976.

34. Vere, D. Controlled clinical trials: the current ethical debate (editorial). *J R Soc Med.* 74:85–88, 1981.

35. Cancer Research Campaign Working Party in Breast Conservation. Informed consent: ethical, legal, and medical implications for doctors and patients who participate in randomized clinical trials. *Br Med J.* 286:1117–1121, 1983.

36. Brewin, T.B. Consent to randomised treatment. *Lancet* ii:919–921, 1982.

37. Wilhelmsen, L. Ethics of clinical trials — the use of placebo (editorial). *Eur J Clin Pharmacol.* 16:295–297, 1979.

38. Bok, S. The ethics of giving placebos. *Sci Am.* 231:17–23, November 1974.

39. Howard, J. and Friedman, L. Protecting the scientific integrity of a clinical trial: some ethical dilemmas. *Clin Pharmacol Ther.* 29:561–569, 1981.

40. Veatch, R.M. Longitudinal studies, sequential design, and grant renewals: what to do with preliminary data. *IRB* 1:1–3, 1979.

41. Friedman, L., and DeMets, D. The data monitoring committee: how it operates and why. *IRB* 3:6–8, 1981.

42. Cassileth, B.R., Zupkis, R.V., Sutton-Smith, K., et al. Informed consent — why are its goals imperfectly realized? *N Engl J Med*. 302:896–900, 1980.

43. Grundner, T.M. On the readability of surgical consent forms. *N Engl J Med*. 302:900–902, 1980.

44. Howard, J.M., DeMets, D., and the BHAT Research Group. How informed is informed consent? The BHAT experience. *Controlled Clin Trials* 2:287–303, 1981.

45. Cassileth, B.R., Lusk, E.J., Miller, D.S., et al. Attitudes toward clinical trials among patients and the public. *JAMA* 248:968–970, 1982.

CHAPTER 2

What Is The Question?

The planning of a clinical trial depends on the question that the investigator is addressing. The general objective is usually obvious, but the specific question to be answered by the trial is often not stated. Stating the question clearly and in advance encourages proper design. It also enhances the credibility of the findings. One would like answers to a number of questions, but the study should be designed with only one major question in mind. This chapter will discuss the selection of this primary question and appropriate ways of answering it. In addition, types of subsidiary questions will be reviewed.

FUNDAMENTAL POINT

Each clinical trial must have a primary question. The primary question, as well as any subsidiary questions, should be carefully selected, clearly defined, and stated in advance.

SELECTION OF THE QUESTIONS

The *primary question* should be the one the investigators are most interested in answering and which is capable of being adequately answered. It is the question upon which the sample size of the study is based, and which must be emphasized in any reporting of the trial results. The primary question may be framed in the form of testing a hypothesis,[1] because most of the time an intervention is postulated to have a particular outcome which, on the average, will be different from the outcome in a control group. The outcome may be a beneficial action such as saving a life, ameliorating an illness, or improving quality of life, or the modification of a characteristic such as behavior, severity of pain or blood pressure.

In some instances, however, the investigator may be interested in demonstrating no difference between intervention and control or between two interventions. For example, are people equally as well off if treated at home for a myocardial infarction as they are if hospitalized? This particular question was addressed recently.[2] Proving that responses to different interventions are the same on the average is not straightforward, and is not the same as failing to demonstrate that a difference exists between the groups. Such proof requires special attention to statistical power (sensitivity); that is, the ability to show a difference had one truly been present.

There may also be a variety of subsidiary, or *secondary questions* related to the primary question. The study may be designed specifically to help address these, or else data collected for the purpose of answering the primary question may also elucidate the secondary questions. They can be of two types. In the first, the response variable is different than that in the primary question. For example, the primary question might ask whether mortality from any cause is altered by the intervention. Secondary questions might relate to incidence of cause-specific death (such as coronary heart disease mortality), sex or age-specific mortality, incidence of non-fatal myocardial infarction, or incidence of stroke.

The second type of secondary question is a *subgroup hypothesis*. For example, in a study of cancer therapy, the investigator may want to look specifically at people by stage of disease at entry into the trial. Such people in the intervention group can be compared with similar people in the control group. Subgroup hypotheses should be (1) specified before data collection begins, (2) based on reasonable expectations, and (3) limited in number. In any event, the number of subjects in any subgroup is usually too small to prove or disprove a subgroup hypothesis.

Both types of secondary questions raise several methodologic issues; for example, if enough statistical tests are done, a few will be significant by chance alone when there is no true intervention effect. Therefore, when a number of tests are carried out, results should be interpreted cautiously. Shedding light or raising new hypotheses is a more likely outcome of these analyses than are conclusive answers. See Chapter 16 for further discussion.

Both primary and secondary questions should be important and relevant scientifically, medically, or for public health purposes. Subject safety and well-being must always be considered in evaluating importance. As reviewed in Chapter 1, potential benefit and risk of harm should be looked at by the investigator, as well as by local Institutional Review Boards (Human Experimentation Committees).

INTERVENTION

When the question is conceived, investigators, at the very least have in mind a class or type of intervention. More commonly, they know the precise drug or procedure they wish to study. In reaching such a decision, they need to consider several aspects. First, the potential benefit of the intervention must be maximized, while possible toxicity is kept to a minimum. Thus, dose of drug or intensity of rehabilitation and frequency of administration are key factors that need to be determined. Investigators must also decide whether to use a single drug, fixed or adjustable doses, sequential drugs, or drug combinations. The composition of the control group regimen is an additional factor.

Second, the availability of the drug for testing needs to be determined. If it is not yet licensed, special approval from the regulatory agency and cooperation by the manufacturer are required.

Third, investigators must take into account design aspects, such as time of initiation and duration of the intervention, and the logistics of blinding.

RESPONSE VARIABLES

Response variables are outcomes measured during the course of the trial, and they define and answer the questions. A response variable may be total mortality, death from a specific cause, incidence of a disease, a complication or specific adverse effect of a disease, symptomatic relief, a clinical finding, or a laboratory measurement. If the primary question concerns total mortality, the occurrence of deaths in the trial clearly answers the question. If the primary question involves severity of arthritis, on the other hand, extent of mobility or a measure of freedom from pain may be reasonably good indicators. In other circumstances, a specific response variable may only partially reflect the overall question. As seen from the above examples, the response variable may show a change from one discrete state (living) to another (dead), from one discrete state to any of several other states (changing from one stage of disease to another) or from one level of a continuous variable to another. If the question can be appropriately defined using a continuous variable, the required sample size may be reduced (Chapter 7). However, the investigator needs to be careful that this variable and any observed differences are clinically meaningful and that the use of a continuous variable is not simply a device to reduce sample size.

In general, a single response variable should be identified to answer the primary question. If more than one are used, the probability of getting a nominally significant result by chance alone is increased (Chapter 16). In addition, if several response variables give inconsistent results, interpretation becomes difficult. The investigator would then need to consider which outcome is most important, and explain why the others gave conflicting results. Unless he has made the determination of relative importance prior to data collection, his explanations are likely to be unconvincing.

Although the practice is not advocated, there may be times when more than one "primary" response variable needs to be looked at. This may be the case when an investigator truly cannot state which of several response variables relates most closely to the primary question. Ideally, the trial would be postponed until this decision can be made. However, overriding concerns, such as increasing use of the intervention in general medical practice, may compel him to conduct the study earlier. In these circumstances, rather than arbitrarily selecting one response variable which may, in retrospect, turn out to be inappropriate, investigators prefer to list several "primary" outcomes. For instance, in the Urokinase Pulmonary Embolism Trial[3] lung scan, arteriogram and hemodynamic measures were given as the "primary" response variables in assessing the effectiveness of the agents urokinase and streptokinase. Chapter 7 discusses the calculation of sample size when a study with several "primary" response variables is designed.

Combining events to make up a response variable might be useful if any one event occurs too infrequently for the investigator reasonably to expect a significant difference without using a large number of subjects. The combined events should be capable of meaningful interpretation. In answering a question where the response variable involves a combination of events, only *one event per subject* should be counted.

One kind of combination response variable involves two kinds of events. In a study of heart disease, combined events might be death from coronary heart disease plus nonfatal myocardial infarction. This is clinically meaningful since death from coronary heart disease and non-fatal myocardial infarction might together represent a measure of coronary heart disease. Difficulties in interpretation can arise if the results of each of the components in such a response variable are inconsistent. When this kind of combination response variable is used, the rules for establishing a hierarchy of events should be established in advance. Thus, a fatal event would take precedence over a nonfatal event, and only the fatal event would be counted.

Another kind of combination response variable involves multiple events of the same sort. Rather than simply asking whether an event has occurred, the investigator can look at the frequency with which it occurs. This may be a more meaningful way of looking at the question than seeking a yes-no outcome. For example, frequency of recurrent transient ischemic attacks or epileptic seizures within a specific follow-up period might comprise the primary response variable of interest. Simply adding up the number of recurrent episodes and dividing by the number of subjects in each group in order to arrive at an average is improper. Multiple events in an individual are not independent, and averaging gives undue weight to those subjects with more than one episode. One approach is to compare the number of subjects with none, one, two, etc episodes; that is, the distribution, by individual, of the number of episodes.

Regardless of whether an investigator is measuring a primary or secondary response variable, certain rules apply. First, he should define and write the questions in advance, being as specific as possible. He should not simply ask, "Is A better than B?" Rather, he should ask, "In population W is drug A at daily dose X more efficacious in reducing Z over a period of time T than drug B at daily dose Y?" Implicit here is the magnitude of the difference that the investigator is interested in detecting. Stating the questions and response variables in advance is essential for planning of study design and calculation of sample size. As shown in Chapter 7, sample size calculation requires specification of the response variables as well as estimates of the effect of intervention. In addition, the investigator is forced to consider what he means by a successful intervention. For example, does the intervention need to reduce mortality by 10% or 25% before a recommendation for its general use is made? Since such recommendations also depend on the frequency and severity of adverse effects, a successful result cannot be completely defined beforehand. However, if a 10% reduction in mortality is clinically important, that should be stated, since it has sample size implications. Specifying response variables and anticipated benefit in advance also eliminates the possibility of the legitimate criticism that can be made if the investigator looked at the data until he found a statistically significant result and then decided that *that* response variable was what he really had in mind all the time.

Second, the primary response variable must be capable of being assessed in all subjects. Selecting one response variable to answer the primary question in some subjects, and another response variable to answer the same primary question in other subjects is not a legitimate practice. It implies that each response variable answers the question of interest with the same precision and accuracy; that each measures exactly the same thing. Such agreement is unlikely. Similarly, response variables should be measured in the same way for all subjects. Measuring a given variable by different instruments or techniques implies that the instruments or techniques yield precisely the same information. This rarely, if ever, occurs. If response variables can be measured only one way in some subjects and another way in other subjects, two separate studies are actually being performed, each of which is likely to be too small.

Third, unless there is a combination primary response variable in which the subject remains at risk of having additional events, a subject's participation generally ends when the primary response variable occurs.

"Generally" is used here because, unless death is the primary response variable, the investigator may well be interested in certain events subsequent to the occurrence of the primary response variable. These events will not change the analysis of the primary response variable but may affect the interpretation of results. For example, deaths taking place after a nonfatal primary response variable has already occurred, but before the official end of the trial as a whole, may be of interest. On the other hand, if a secondary response variable occurs, the subject should remain in the study (unless, of course, it is a fatal secondary response variable). He must continue to be followed because he is still at risk of developing the primary response variable. A study of heart disease may have, as its primary question, death from coronary heart disease and, as a secondary question, incidence of nonfatal myocardial infarction. If a subject suffers a nonfatal myocardial infarction, this counts toward the secondary response variable. However, he must remain in the study for analytic purposes and be at risk of dying (the primary response variable). This is true whether or not he is continued on the intervention regimen. If he does not remain in the study for purposes of analysis of the primary response variable, bias may result. (See Chapter 16 for more discussion of subject withdrawal.)

Fourth, response variables should be capable of unbiased assessment. Truly double-blind studies have a distinct advantage over other studies in this regard. If a trial is not double-blind (Chapter 6), then, whenever possible, response variable assessment should be done by people who are not participating in subject follow-up and who are blinded to the identity of the study group. Independent reviewers are often helpful. Of course, the use of blinded or independent reviewers does not solve the problem of bias. Unblinded investigators sometimes fill out forms and the subjects may be influenced by the investigators. This may be the case during an exercise performance test, where the impact of the person administering the test on the results may be considerable. Some studies arrange to have the intervention administered by one investigator and response variables evaluated by another. Unless the subject is blinded (or otherwise unable to communicate), this procedure is also vulnerable to bias. One solution to this dilemma is to use only "hard," or objective, response variables (which are unambiguous and not open to interpretation). This assumes complete and honest ascertainment of outcome. Double-blind studies have the advantage of allowing the use of softer response variables, since the risk of assessment bias is minimized.

Fifth, it is important to have response variables that can be ascertained as completely as possible. A hazard of long-term studies is that subjects may fail to return for follow-up appointments. If the response variable is one that depends on an interview or an examination, and subjects fail to return for follow-up appointments information will be lost. Not only will it be lost, but it may be differentially lost in the intervention and control groups. When appropriate, death is a useful response variable in these circumstances because the investigator can usually ascertain vital status, even if the subject is no longer actively participating in the study.

A common criticism of clinical trials is that they are expensive and of long duration. This is particularly true for trials which use the occurrence of clinical events as the primary response variable. It has been suggested that response variables which are continuous in nature might substitute for the clinical outcomes. Thus, instead of monitoring cardiovascular mortality or myocardial infarction an investigator could examine progress of atherosclerosis by means of angiography, or change in cardiac arrhythmia by means of ambulatory electrocardiograms or programmed electrical stimulation. In the cancer field, change in tumor size might replace mortality.

An argument for use of these "surrogate response variables" is that since they are continuous, the sample size can be smaller than otherwise. In addition, changes in these variables are likely to occur before the clinical event, shortening the time required for the trial.

These are valid objectives. Before using surrogate response variables, however, an investigator needs to consider several issues. First, does the variable truly reflect the clinical outcome? Have changes in the surrogate variable been shown to be highly correlated with changes in the clinical variable? Second, can the surrogate variable be assessed accurately and reliably? Is there so much measurement error that in fact the sample size increases or the results are questioned? Third, will the evaluation be so unacceptable to the subject that the study will become infeasible? If it requires invasive techniques, subjects may refuse to join the trial, or worse, discontinue participation before the end. Measurement can require expensive equipment and highly trained staff, which may, in the end, make the trial more costly than if clinical events are monitored. Finally, will the conclusions of the trial be accepted by the scientific and medical communities? If there is insufficient acceptance that the surrogate variable reflects clinical outcome, inspite of the investigator's conviction, there is little point in using such variables.

All clinical trials are compromises between the ideal and the practical. This is true in the selection of primary response variables. The most objective or those most easily measured may either occur too infrequently or fail to define adequately the primary question. To select a response variable which can be reasonably and reliably assessed and yet which can provide an answer to the primary question requires judgment. If such a response variable cannot be found, the wisdom of conducting the trial should be re-evaluated.

ADVERSE EFFECTS

Important questions that can be answered by clinical trials concern adverse or side reactions to therapy (Chapter 11). Here, unlike the primary or secondary questions, it is not always possible to specify in advance the question to be answered. What adverse reactions might occur, and their severity, may be unpredictable. Furthermore, rigorous, convincing demonstration of serious toxicity is usually not achieved, because it is generally thought unethical to continue a study to the point at which a drug has been conclusively shown to be more harmful than beneficial.[4-6] Investigators traditionally monitor a variety of laboratory and clinical measurements, look for possible adverse effects, and compare these in the intervention and control groups. Statistical significance and the previously mentioned problem of multiple response variables become secondary to clinical judgment and subject safety. While this will lead to the conclusion that some purely chance findings are labeled as adverse effects, moral responsibility to the subjects requires a conservative attitude toward safety monitoring, particularly if an alternative therapy is available.

NATURAL HISTORY

Though it is not intervention-related, an often valuable use of the collected data, especially in long-term trials, is a natural history study in the control group. If the control group is either on placebo or no systematic treatment, various baseline factors may be studied for their relation to specific outcomes. Assessment of the prognostic importance of these factors can lead to better understanding of the disease under study and development of new hypotheses. Of course, generally only predictive association—and not necessarily causation—may properly be inferred from such data. The study subjects may be a highly selected group and natural history findings must be interpreted in that light.

Since they are not study hypotheses, specific natural history questions need not be specified in advance. However, properly designed baseline forms require some advance consideration of which factors might be related to outcome. After the study has started, going back to ascertain missing baseline information in order to answer natural history questions is generally a fruitless pursuit. At the same time, collecting large amounts of baseline data on the slight chance that they might provide useful information costs money, consumes valuable time, and may lead to less careful collection of important data. It is better to restrict data collection to those baseline factors that are known, or seriously thought, to have an impact on prognosis.

ANCILLARY QUESTIONS

Often a clinical trial can be used to answer questions which do not bear directly on the intervention being tested, but which are nevertheless of interest. The structure of the trial and the ready access to subjects may make it the ideal vehicle for such investigations. Weinblatt, Ruberman, and colleagues reported that a low level of education among survivors of a myocardial infarction was a marker of poor risk of survival.[7] The authors subsequently evaluated whether the educational level was an indicator of psychosocial stress.[8] To further investigate these findings, they performed a study ancillary to the Beta-Blocker Heart Attack Trial,[9] which evaluated whether the regular administration of propranolol could reduce three-year mortality in people with acute myocardial infarctions. Interviews assessing factors such as social interaction, attitudes, and personality were conducted in over 2300 men in the trial.[10] Inability to cope with high life stress and social isolation were found to be significantly and independently associated with mortality. Effects of low education were accounted for by these two factors. By enabling the investigators to perform this study, the Beta-Blocker Heart Attack Trial provided an opportunity to examine an important issue in a large sample, even though it was peripheral to the main question.

GENERAL COMMENTS

Although this text attempts to provide straightforward concepts concerning the selection of study response variables, things are rarely as simple as one would like them to be. Investigators often encounter problems related to design, data monitoring and ethical issues and interpretation of study results.

In long-term studies of high-risk subjects, where total mortality is not the primary response variable, many subjects may nevertheless die. They are, therefore, removed from the population at risk of developing the response variable of interest. Even in relatively short studies, if the subjects are seriously ill, death may occur. In designing studies, therefore, if the primary response variable is either a nonfatal event or cause-specific mortality, the investigator needs to consider the impact of total mortality for two reasons. First, it will reduce the effective sample size. One would like to allow for this reduction by estimating the overall mortality and increasing sample size accordingly. However, a methodology for estimating mortality and increasing sample size is not yet well defined. Second, if mortality is related to the intervention, either favorably or unfavorably, excluding from study analysis those who die may bias results for the primary response variable.

One solution, whenever the risk of mortality is high, is to choose total mortality as the primary response variable. Alternatively, the investigator can combine total mortality with a pertinent nonfatal event as a combined primary response variable. Neither of these solutions may be appropriate and, in that case, the investigator should monitor total mortality as well as the primary response variable. Evaluation of the primary response variable will then need to consider those who died during the study.

Regardless of whether or not it is the primary response variable, total mortality—as well as any other adverse occurrence—needs to be monitored during a study (Chapter 15). The ethics of continuing a study which, despite a favorable trend for the primary response variable, shows equivocal or even negative results for secondary response variables, or the presence of major adverse effects, are questionable. Deciding what to do is difficult if an intervention is giving promising results with regard to death from a specific cause, (which may be the primary response variable), yet total mortality is unchanged or increased. This issue arose in a study of clofibrate in people with elevated serum cholesterol.[11]

Finally, conclusions from data are not always clear-cut. Issues such as quality of life or annoying long-term side effects may cloud results that are clear with regard to primary response variables such as increased survival. In such circumstances, the investigator must offer his best assessment of the results but should report sufficient detail about the study to permit others to reach their own conclusions (Chapter 18).

REFERENCES

1. Cutler, S.J., Greenhouse, S.W., Cornfield, J., and Schneiderman, M.A. The role of hypothesis testing in clinical trials: biometrics seminar. *J Chronic Dis.* 19:857–882, 1966.

2. Hill, J.D., Hampton, J.R., and Mitchell, J.R.A. A randomised trial of home-versus-hospital management for patients with suspected myocardial infarction. *Lancet* 1:837–841, 1978.

3. Urokinase Pulmonary Embolism Trial Study Group. Urokinase Pulmonary Embolism Trial: phase 1 results. *JAMA* 214:2163–2172, 1970.

4. Shimkin, M.B. The problem of experimentation on human beings. I. The research worker's point of view. *Science* 117:205–207, 1953.

5. Chalmers, T.C. Invited Remarks: National Conference on Clinical Trials Methodology. *Clin Pharmacol Ther.* 25:649–650, 1979.

6. Stamler, J. Invited Remarks: National Conference on Clinical Trials Methodology. *Clin Pharmacol Ther.* 25:651–654, 1979.

7. Weinblatt, E., Ruberman, W., Goldberg, J.D., et al. Relation of education to sudden death after myocardial infarction. *N Engl J Med.* 299:60–65, 1978.

8. Ruberman, W., Weinblatt, E., Goldberg, J.D., et al. Education, psychosocial stress, and sudden cardiac death. *J Chron Dis.* 36:151–160, 1983.

9. Beta-Blocker Heart Attack Trial Research Group. A randomized trial of propranolol in patients with acute myocardial infarction. I. Mortality results. *JAMA*. 247:1707–1714, 1982.

10. Ruberman, W., Weinblatt, E., Goldberg, J.D., et al. Psychosocial influences on mortality after myocardial infarction. *N Engl J Med*. 311:552–559, 1984.

11. Report from the Committee of Principal Investigators. A co-operative trial in the primary prevention of ischaemic heart disease using clofibrate. *Br Heart J*. 40:1069–1118, 1978.

CHAPTER 3

Study Population

Defining the study population is an integral part of posing the primary question. It is not enough to claim that an intervention is or is not effective without describing the type of subject on which the intervention was tested. The description requires specification of criteria for subject eligibility. This chapter focuses on how to define the study population. In addition, it considers two questions. First, to what extent will the results of the trial be generalizable to a broader population? Second, what impact does selection of eligibility criteria have on subject recruitment, or, more generally, study feasibility?

FUNDAMENTAL POINT

The study population should be defined in advance, stating unambiguous inclusion (eligibility) criteria. The impact that these criteria will have on study design, ability to generalize, and subject recruitment must be taken into account.

DEFINITION OF STUDY POPULATION

The study population is the subset of the general population defined by the eligibility criteria. The group of subjects actually studied in the trial is selected from the study population.

In reporting the study, the investigator needs to say what people were studied and how they were selected. The reasons for this are several. Most important, if an intervention is shown to be successful or unsuccessful, the medical and scientific communities must know to what kinds of people the findings apply.

Second, knowledge of the study population helps other investigators to assess the study's merit and appropriateness. For example, an antianginal drug may be found to be ineffective. Close examination of the description of the study population, however, could reveal that the subjects represented a variety of ill-defined conditions characterized by chest pain. Thus, the study may not have been properly designed to evaluate the antianginal effects of the agent. Since many literature reports contain inadequate characterization of the study subjects, readers are often unable fully to assess the merit of the studies. Third, in order for other investigators to be able to replicate the study, they need data descriptive of the subjects enrolled. A similar issue is sometimes found in laboratory research. Because of incomplete discussion of details of the methodology, procedures and preparation of materials, other investigators find it impossible to replicate an experiment. Before most research findings are widely accepted, they need to be confirmed by independent scientists. Only small trials are likely to be repeated, but these are the ones, in general, that most need confirmation.

Inclusion criteria and reasons for their selection should be stated in advance. Ideally, all eligibility criteria should be precisely specified, but this is often impractical. Therefore, those criteria central to the study should be the most carefully defined. For example, in a study of survivors of a myocardial infarction, the investigator may be interested in excluding people with severe hypertension. He will require an explicit definition of myocardial infarction; but with regard to hypertension, it may be sufficient to state that people with a diastolic blood pressure above a specified level will be excluded. Note that even here, the definition of severe hypertension, though arbitrary, is fairly specific. In a study of antihypertensive agents, however, the above definition of severe hypertension is inadequate. If the investigator wants to include only people with diastolic blood pressure over 90 mm Hg, he should specify how often it is to be determined, when, with what instrument, by whom, and in what circumstances. If the etiology of the blood pressure is also important (eg, he wants only people with hypertension secondary to pheochromocytoma), he needs to specify the etiology and how it is to be determined. For any study of antihypertensive agents, the criterion of hypertension is central; a detailed definition of myocardial infarction, on the other hand, may be less important.

If age is a restriction, the investigator should specify not only that a subject must be less than 70, for example, but *when* he must be less than 70. If a subject is 69 at the time of a pre-baseline screening examination, but 70 at baseline, is he eligible? This should be clearly indicated. If diabetes mellitus is an exclusion criterion, is this only insulin-dependent diabetes, only juvenile-onset diabetes, or all diabetes? Does chemical diabetes warrant exclusion? How is diabetes mellitus defined? Often there are no "correct" ways of defining inclusion and exclusion criteria — arbitrary decisions must be made. Regardless, they need to be as clear as possible, with complete specifications of the technique and laboratory methods.

In general, eligibility criteria relate to subject safety and study design. The following categories outline the framework upon which to develop individual criteria:

1. Subjects who have the potential to benefit from the intervention are obviously candidates for enrollment into the study. The investigator selects subjects on the basis of his scientific knowledge and the expectation that the intervention will work in a specific way on a certain kind of subject. For example, subjects with a genito-urinary coliform infection are appropriate to enroll in a study of a new antibiotic agent known to be effective in vivo and thought to penetrate to the site of the infection in sufficient concentration. It should be evident from this example that selection of the subject depends on knowledge of the mechanism of action of the intervention. Knowing the mechanism of action may enable the investigator to identify a well-defined group of subjects likely to respond to the intervention. Thus, people with similar characteristics with respect to the relevant variable, that is, a *homogeneous* population, can be studied. In the above example, subjects are homogeneous with regard to the type and strain of bacteria, and to site of infection. If age or renal or liver function are also critical, these too might be considered, creating an even more highly selected group.

Even if the mechanism of action of the intervention is known, however, it may not be feasible to identify a homogeneous population because the technology to do so may not be available. For instance, the causes of headache are numerous and, with few exceptions, not easily or objectively determined. If a potential therapy were developed for one kind of headache, it would be difficult to identify precisely the people who might benefit.

If the mechanism of action of the intervention is unclear, or if there is uncertainty at which stage of a disease a treatment might be most beneficial, a specific group of subjects likely to respond cannot easily be selected. The Diabetic Retinopathy Study[1] evaluated the effects of photocoagulation. In this trial, each subject had one eye treated while the other eye served as the control. Subjects were subgrouped on the basis of existence, location and severity of vessel proliferation. Before the trial was scheduled to end, it became apparent that treatment was dramatically effective in the four most severe of the ten subgroups. To have initially selected for study only those four subgroups who benefited was not possible given existing knowledge.

Some interventions may have more than one potentially beneficial mechanism of action. For example, if exercise reduces mortality or morbidity (an as yet uncertain conclusion), is it because of its effect on cardiac performance, its weight-reducing effect, its effect on the subject's sense of well being, some combination of these effects, or some as yet unknown effect? The investigator could select study subjects who have poor cardiac performance, or who are obese or who, in general, do not feel well. If he chose incorrectly, his study would not yield a positive result. If he chose subjects with all three characteristics and then showed benefit from exercise, he would never know which of the three aspects was important.

One could, of course, choose a study population, the members of which differ in one or more identifiable aspects of the condition being evaluated, or a *heterogeneous* group. These could include stage or severity of a disease, etiology, or demographic factors. In the above example, studying a heterogeneous population may be preferable. By comparing outcome with presence or absence of initial obesity or sense of well being, the investigator may discover the relevant characteristics and gain insight into the mechanism of action. Also, when the study group is too restricted, there is no opportunity to discover whether an intervention is effective in a subgroup not initially considered. The broadness of the Diabetic Retinopathy Study was responsible for showing that the remaining six subgroups also benefited from therapy.[2] If knowledge had been more advanced, only the four subgroups with the most dramatic improvement might have been studied. Obviously, after publication of the results of these four subgroups, another trial might have been initiated. However, valuable time would have been wasted. Possibly inappropriate extrapolation of conclusions to milder retinopathy might even have made a second study impossible. Of course, the effect of the intervention on a heterogeneous group may be diluted. The ability to detect a benefit may be reduced. That is the price to be paid for incomplete knowledge about mechanism of action.

Homogeneity and heterogeneity are matters of degree and knowledge. As scientific knowledge advances, ability to classify is improved. Today's homogeneous group may be considered heterogeneous tomorrow. The discovery of Legionnaires' Disease[3] as a separate entity, caused by a specific organism improved possibilities for categorizing respiratory disease. Presumably, until that discovery, people with Legionnaires' Disease were simply lumped together with people having other respiratory ailments.

2. In selecting subjects to be studied, not only does the investigator require people in whom the intervention might work; he also wants to choose people in whom there is a high likelihood that he can detect the hypothesized results of the intervention. Careful choice will enable investigators to detect results in a reasonable period of time, given a reasonable number of subjects and a finite amount of money.

For example, in a trial of an antiarrhythmic drug, an investigator would not wish to enroll people who, in the past two years, have had only one brief cardiac arrhythmic episode (assuming such a person could be identified). The likelihood of the antiarrhythmic agent's making an impact on the health of this person is limited, since his likelihood of having many future arrhythmic episodes is small. Persons with frequent episodes would be more appropriate. Similarly, many people accept the hypothesis that cholesterol is a continuous variable in its impact on the risk of developing cardiovascular disease. That is, there is no specific level below which the serum cholesterol concentration is good with respect to coronary heart disease. Theoretically, an investigator could take any population with moderate or low cholesterol, attempt to lower its cholesterol, and see if occurrence of cardiovascular disease is reduced. However, this would require studying an impossibly large number of people, since the calculation of sample size (Chapter 7) takes into account expected frequency of the primary response variables. When the expected frequency in the control group is low, as it would likely be in people who do not have elevated serum cholesterol, the number of people studied must be correspondingly high. From a sample size point of view it is, therefore, desirable to begin studying people with a high expected event rate. If results from a first trial are positive, the investigator can then go to groups with lower levels. The initial Veterans Administration study of the treatment of hypertension[4] involved people with diastolic blood pressure from 115 through 129 mm Hg. After therapy was shown to be beneficial in that group, a second trial was undertaken using people with diastolic blood pressures from 90 through 114 mm Hg.[5] The latter study suggested that treatment should be instituted for people with diastolic blood pressure over 104 mm Hg. Results were less clear for people with lower blood pressure. Subsequently, the Hypertension Detection and Follow-up Program[6,7] demonstrated benefit from treatment for people with diastolic blood pressure of 90 mm Hg or above.

Generally, if the primary response is continuous (eg, blood pressure, blood sugar, body weight), change is easier to detect when the initial level is extreme. In a study to see whether a new drug is antihypertensive, one might expect a more pronounced drop of blood pressure in a subject with diastolic pressure of 130 mm Hg than in one with diastolic pressure of 102 mm Hg. There are exceptions to this rule, especially if a condition has multiple causes. The relative frequency of each cause might be different across the spectrum of values. For example, genetic disorders might be heavily represented among subjects with extremely high cholesterol. These lipid disorders may require alternative therapies or may even be resistant to usual methods of reducing cholesterol. In addition, use of subjects with lower levels of a variable such as cholesterol might be less costly.[8] This is because of lower screening costs. Therefore, while in general, use of higher risk subjects is preferable, other considerations can modify this.

3. Most interventions are likely to have adverse effects. The investigator needs to weigh these against possible benefit when he evaluates

the feasibility of doing the study. However, any subject to whom the intervention is known to be harmful should not, except in unusual circumstances, be admitted to the trial. Pregnant women—or those who may become pregnant during the study—are often excluded from drug trials (unless, of course, the primary question concerns pregnancy). The small amount of additional data obtained does not justify the risk of possible teratogenicity. Similarly, investigators would probably exclude from a study of almost any of the anti-inflammatory drugs people with a history of gastric bleeding. Gastric bleeding is a fairly straightforward and absolute contraindication for enrollment. Yet, an exclusion criterion such as "history of major gastric bleed," leaves much to the judgment of the investigator. The wording implies that gastric hemorrhaging is not an absolute contraindication, but a relative one that depends upon clinical judgment. The phrase also recognizes the question of anticipated risk versus benefit, because it does not clearly prohibit people with a mild bleeding episode in the distant past from being placed on an anti-inflammatory drug. It may very well be that such people take aspirin or similar agents—possibly for a good reason—and studying such people may prove more beneficial than hazardous.

Note that these exclusions apply only before enrollment into the trial. During a trial subjects may develop conditions which would have excluded them had any of these conditions been present earlier. In these circumstances, the subject may be removed from the intervention regimen, but he should be kept in the trial for purposes of analysis.

4. Subjects at high risk of developing conditions which preclude the ascertainment of the event of interest should be excluded from enrollment. The intervention may or may not be efficacious in such subjects, but the necessity for excluding them from enrollment relates to design considerations. The issue of competing risk is generally of greater interest in long-term studies. In many studies of people with heart disease, those who have cancer or severe kidney or liver disorders are excluded because these diseases might cause the subject to die or withdraw from the study before the primary response variable can be observed. However, even in short-term studies, the competing risk issue needs to be considered. For example, an investigator may be studying a new intervention for a specific congenital heart defect in infants. Such infants are also likely to have other life-threatening defects. The investigator would not want to enroll infants if one of these other conditions were likely to lead to the death of the infant before he had an opportunity to evaluate the effect of the intervention. (This matter is similar to the one raised in Chapter 2, which presented the problem of the impact of high expected total mortality on a study in which the primary response variable is morbidity or cause-specific mortality.) When there is competing risk, the ability to assess the true impact of the intervention is, at

best, lessened. At worst, if the intervention somehow has either a beneficial or harmful effect on the coexisting condition, biased results for the primary question can be obtained.

5. Investigators prefer, ordinarily, to enroll only subjects who are likely to comply with the study protocol. Subjects are expected to take their assigned intervention (usually a drug) and return for scheduled follow-up appointments regardless of the intervention assignment. In unblinded studies, subjects are asked to accept the random assignment, even after knowing its identity, and abide by the protocol. Moreover, subjects should not receive the study intervention from sources outside the trial during the course of the study. Subjects should also refrain from using other interventions which may compete with the study intervention. Noncompliance by subjects reduces the opportunity to observe the true effect of intervention. Unfortunately, there are no failsafe ways of selecting perfect subjects. Traditional guidelines have led to disappointing results. For a further discussion of compliance, see Chapter 13.

GENERALIZATION

Study subjects are usually nonrandomly chosen from the study population, which in turn is defined by the eligibility criteria. As long as selection of subjects into a trial occurs, they must be regarded as special. Therefore, investigators have the problem of generalizing from subjects actually in the trial to the study population and then to a wider population. Defined medical conditions and quantifiable or discrete variables such as age, sex, or elevated blood sugar can be clearly stated and measured. For these characteristics, specifying in what way the study subjects and study population are different from the general population is relatively easy. Judgments about the appropriateness of generalizing study results can, therefore, be made. Other factors of the study subjects are less easily characterized. Obviously, an investigator studies only those subjects available to him. If he lives in Florida, he will not be studying people living in Maine. Even within a geographical area, most investigators are hospital-based. Except in rare instances, subjects are not drawn from the community at large. Furthermore, many hospitals are referral centers. Only certain types of subjects come to the attention of investigators at these institutions. It may be impossible to decide whether these factors are relevant when generalizing to other geographical areas or medical settings.

It is often forgotten that subjects must agree to enroll in a study. What sort of person volunteers for a study? Why do some agree to participate while others do not? The requirement that study subjects sign informed consent or return for periodic examinations is sufficient to make certain subjects unwilling to participate. The reasons are not often obvious. What is known, however, is that volunteers can be different from non-volunteers.[9,10] They are usually in better health, and they are more likely to comply with the study protocol. However, the reverse could also be true. A person might be more motivated if he has disease symptoms. In the absence of knowing what motivates the particular study subjects, appropriate compensatory adjustments cannot be made. Because specifying how volunteers differ from others is difficult, an investigator cannot confidently identify those segments of the study population or the general population that these study subjects supposedly represent. One approach to answering the question of representativeness is to maintain a log which lists prospective subjects identified, but not enrolled, and the reasons for excluding them. This log can provide an estimate of the proportion of all potentially eligible people who meet study entrance requirements and can also indicate how many otherwise eligible people refused enrollment. In an effort to further assess the issue of representativeness, response variables in those excluded have also been monitored. In the Norwegian Multicenter Study on timolol,[11] people excluded because of contraindication to the study drug or competing risks had a mortality rate twice that of enrolled subjects.

The Coronary Artery Surgery Study included a randomized trial which compared coronary artery bypass surgery against medical therapy and a registry of people eligible for the trial but who refused to participate.[12] The enrolled and not enrolled groups were alike in most identifiable respects. Survival in the subjects randomly assigned to medical care was the same as those receiving medical care but not in the trial. The findings for those undergoing surgery were similar. Therefore, in this particular case, the trial subjects appeared to be representative of the study population.

Since the investigator can describe only to a limited extent the kinds of subjects in whom an intervention was evaluated, a leap of faith is always required when applying any study findings to the population at large. In taking this jump, one must always strike a balance between making unjustifiably broad generalizations and being too conservative in one's claims. Some extrapolations are reasonable and justifiable from a clinical point of view. The Coronary Drug Project, a trial of lipid-lowering drugs, studied men between the ages of 30 and 64 who had had at least one myocardial infarction.[13] Conclusions from that trial might reasonably be extended to men below 30 and above age 64, since conclusions about younger men in the trial were similar to conclusions about older men.[14] Here too, though, caution should be exercised; causes of myocardial infarctions in young men might be different from causes in older men. One might be tempted to claim that the findings could be extrapolated to women, as men were chosen mainly for statistical reasons (greater incidence of events related to coronary heart disease). An extrapolation from men to women could be hazardous because there was no information at all on women. Furthermore, without supporting evidence, the claim that results from the Coronary Drug Project could apply to people who had not yet suffered a heart attack or to other methods of lowering lipids, such as diet, are probably inappropriate.

RECRUITMENT

The impact of eligibility criteria on recruitment of patients should be considered when deciding on these criteria. Using excessive restrictions in an effort to obtain a pure (or homogeneous) sample can lead to extreme difficulty in getting sufficient subjects. Age and sex are two criteria which have obvious bearing on the ease of enrolling subjects. The Coronary Prevention Trial undertaken by the Lipid Research Clinics was a collaborative trial evaluating a lipid-lowering drug in men between the ages of 35 and 59 with severe hypercholesterolemia. One of the Lipid Research Clinics[15] noted that approximately 35,000 people were screened and only 257 subjects enrolled. Exclusion criteria, all of which were perfectly reasonable and scientifically sound, coupled with the number of people who refused to enter the study, brought the yield down to less than 1%. As discussed in Chapter 9, this example of greater-than-expected numbers being screened — as well as unanticipated problems in reaching potential subjects — is common to most clinical trials.

If entrance criteria are properly determined in the beginning of a study, there should be no need to change them. As discussed earlier in this chapter, eligibility criteria are appropriate if they exclude those who might be harmed by the intervention, those who are not likely to be benefited by the intervention, or those who are not likely to comply with the study protocol. The reasons for each criterion should be carefully examined during the planning phase of the study. If they do not fall into one of the above categories, they should be reassessed. Whenever an investigator considers changing criteria, he needs to look at the effect of changes on subject safety and study design. It may be that, in opening the gates to accommodate more subjects, he increases the required sample size, because the subjects admitted may have lower probability of developing the primary response variable. He can thus lose the benefits of added recruitment. In summary, capacity to recruit patients and effectively to carry out the trial could greatly depend on the eligibility criteria that are set. As a consequence, careful thought should go into establishing them.

REFERENCES

1. Diabetic Retinopathy Study Research Group: Preliminary report on effects of photocoagulation therapy. *Am J Ophthalmol.* 81:383–396, 1976.

2. Diabetic Retinopathy Study Research Group. Photocoagulation treatment of proliferative diabetic retinopathy: the second report of diabetic retinopathy study findings. *Ophthalmol* (formerly *Trans Am Acad Ophthalmol Otolaryngol.)* 85:82–105, 1978.

3. Fraser, D.W., Tsai, T.R., Orenstein, W. et al. Legionnaires' Disease: description of an epidemic of pneumonia. *N Engl J Med.* 297:1189–1197, 1977.

4. Veterans Administration Cooperative Study Group on Antihypertensive Agents. Effects of treatment on morbidity in hypertension: results in patients with diastolic blood pressures averaging 115 through 129 mm Hg. *JAMA* 202:1028–1034, 1967.

5. Veterans Administration Cooperative Study Group on Antihypertensive Agents. Effects of treatment on morbidity in hypertension: II. Results in patients with diastolic blood pressure averaging 90 through 114 mm Hg. *JAMA* 213:1143–1152, 1970.

6. Hypertension Detection and Follow-up Program Cooperative Group. The Hypertension Detection and Follow-up Program. *Prev Med.* 5:207–215, 1976.

7. Hypertension Detection and Follow-up Program Cooperative Group. Five-year findings of the Hypertension Detection and Follow-up Program. 1. Reduction in mortality of persons with high blood pressure, including mild hypertension. *JAMA* 242:2562–2571, 1979.

8. Sondik, E.J., Brown, B.W., Jr., and Silvers, A. High risk subjects and the cost of large field trials. *J Chronic Dis.* 27:177–187, 1974.

9. Horwitz, D., and Wilbeck, E. Effect of tuberculosis infection on mortality risk. *Am Rev Respir Dis.* 104:643–655, 1971.

10. Wilhelmsen, L., Ljungberg, S., Wedel, H., and Werko, L. A comparison between participants and non-participants in a primary preventive trial. *J Chronic Dis.* 29:331–339, 1976.

11. Pedersen, T.R. The Norwegian Multicenter Study of timolol after myocardial infarction. *Circulation* 67 (suppl I):I-49–I-53, 1983.

12. CASS Principal Investigators and Their Associates. Coronary Artery Surgery Study (CASS): a randomized trial of coronary artery bypass surgery. Comparability of entry characteristics and survival in randomized patients and nonrandomized patients meeting randomization criteria. *JACC* 3:114–128, 1984.

13. Coronary Drug Project Research Group. The Coronary Drug Project: design, methods, and baseline results. *Circulation* 47 (suppl 1):1–50, 1973.

14. Coronary Drug Project Research Group. Clofibrate and niacin in coronary heart disease. *JAMA* 231:360–381, 1975.

15. Benedict, G.W. LRC Coronary Prevention Trial: Baltimore. *Clin Pharmacol Ther.* 25:685–687, 1979.

CHAPTER 4

Basic Study Design

The foundations for the design of controlled experiments were established for agricultural application. They are described in several classical statistics textbooks.[1-4] From these sources evolved the basic design of controlled clinical studies.

Although the history of clinical experimentation contains several instances in which the need for control groups has been recognized,[5,6] this need was not widely accepted until the 1950s.[7] In the past, when a new intervention was first investigated, it was likely to be given to only a small number of subjects, and the outcome compared, if at all, to that in subjects previously treated in a different manner. The comparison was informal and frequently based on memory alone. Subjects were often evaluated initially and then reexamined after an intervention had been introduced. In such studies, the changes from the initial state were used as the measure of success or failure of the new intervention. What could not be known was whether the subject would have responded in the same manner if there had been no intervention at all. However, then—and unfortunately sometimes even today—this kind of observation has formed the basis for the widespread use of new interventions.

Of course, some results are so highly dramatic that no comparison group is needed. Successful results of this magnitude, however, are rare.

One example is the effectiveness of penicillin in pneumococcal pneumonia. Another example originated with Pasteur who in 1884 was able to demonstrate that a series of vaccine injections protected dogs from rabies.[8] He suggested that due to the long incubation time, prompt vaccination of a human being after infection might prevent the fatal disease. The first patient was a 9-year-old boy who had been bitten three days earlier by a rabid dog. The treatment was completely effective. The confirmation came from another boy who was treated within six days of having been bitten. During the next few years hundreds of patients were given the anti-rabies vaccine. If given within certain time-limits, it was almost always effective.

Gocke reported on a similar, uncontrolled study of patients with acute fulminant viral hepatitis.[9] Nine consecutive cases had recently been observed, all of whom had a fatal outcome. The next diagnosed case, a young staff nurse in hepatic coma, was given immunotherapy in addition to standard treatment. The patient survived as did four others among eight given the antiserum. The author initially thought that this uncontrolled study was conclusive. However, in considering other explanations for the encouraging findings, he could not eliminate the possibility that a tendency to treat patients earlier in the course and more intensive care might be responsible for the observed outcome. Thus, he joined a double-blind, randomized trial comparing hyperimmune anti-Australia globulin to normal human serum globulin in patients with severe acute hepatitis. Nineteen of 28 patients (67.9%) randomized to control treatment died, compared to 16 of 25 patients (64%) randomized to treatment with exogenous antibody, a statistically nonsignificant difference.[10]

A number of medical conditions are either of short duration or episodic in nature. Evaluation of therapy in these cases can be difficult in the absence of controlled studies. Snow and Kimmelman reviewed various uncontrolled studies of surgical procedures for Meniere's disease.[11] They found that about 75% of patients improved, but noted that this is similar to the 70% remission rate occurring naturally.

Given the wide spectrum of the natural history of almost any disease and the variability of an individual subject's response to an intervention, most investigators recognize the need for a defined control or comparison group.

FUNDAMENTAL POINT

Sound scientific clinical investigation almost always demands that a control group be used against which the new intervention can be compared. Randomization is the preferred way of assigning subjects to control and intervention groups.

Statistics and epidemiology textbooks and papers[12-21] present various study designs in some detail. This chapter will discuss several major clinical trial designs: randomized, non-randomized concurrent, historical, cross-over, withdrawal, factorial, group allocation, and combination. The randomized control and the non-randomized concurrent control studies both assign subjects to either the intervention or the control group, but only the former makes the assignment by using a randomized procedure. Historical control studies compare a group of subjects on a new therapy or intervention with a previous group of subjects on standard or control therapy. The cross-over design uses each subject twice, once as a member of the control group and once as a member of the intervention group. Questions have been raised concerning the method of selection of the control group, but the major controversy revolves around the use of historical versus randomized controls.[22-24] Each of the designs has advantages and disadvantages, but a randomized control design is the standard by which other studies should be judged. A discussion of sequential designs is postponed until Chapter 15 because the basic feature involves interim analyses.

For each of the designs it is assumed, for simplicity of discussion, that a single control group and a single intervention group are being considered. These designs can be extended to more than one intervention group and more than one control group.

RANDOMIZED CONTROL STUDIES

The randomized control study is the standard against which all other designs must be compared. Randomized control studies are comparative studies with an intervention group and a control group; the assignment of the subject to a group is determined by the formal procedure of randomization. Randomization, in the simplest case, is a process by which all subjects are equally likely to be assigned to either the intervention group or the control group. The features of this technique are discussed in Chapter 5. There are three advantages of the randomized design over other methods for selecting controls.[24]

First, randomization removes the potential of bias in the allocation of subjects to the intervention group or to the control group. Such allocation bias could easily occur, and cannot be necessarily prevented, in the non-randomized concurrent or historical control study because the investigator or the subject may influence the choice of intervention. This influence can be conscious or subconscious and can be due to numerous factors, including the prognosis of the subject. The direction of the allocation bias may go either way and can easily invalidate the comparison.

The second advantage, somewhat related to the first, is that randomization tends to produce comparable groups; that is, the measured or unknown prognostic factors and other characteristics of the subjects at the time of randomization will be, on the average, evenly balanced between the intervention and control groups. This does not mean that in any single experiment all such characteristics, sometimes called baseline variables or covariates, will be perfectly balanced between the two groups. However, it does mean that for independent covariates, whatever the detected or undetected differences that exist between the groups, the overall magnitude and direction of the differences will tend to be equally divided between the two groups. Of course, many covariates are strongly associated; thus, any imbalance in one would tend to produce imbalances in the others. As discussed in Chapters 5 and 16, stratified randomization and stratified analysis are methods commonly used to guard against and adjust for imbalanced randomizations.

The third advantage of randomization is that the validity of statistical tests of significance is guaranteed. As has been stated,[24] "although groups compared are never perfectly balanced for important covariates in any single experiment, the process of randomization makes it possible to ascribe a probability distribution to the difference in outcome between treatment groups receiving equally effective treatments and thus to assign significance levels to observed differences." The validity of the statistical tests of significance is not dependent on the balance of the prognostic factors between the two groups. The chi-square test for two by two tables and Student's t-test for comparing two means can be justified on the basis of randomization alone without making further assumptions concerning the distribution of baseline variables. If randomization is not used, further assumptions concerning the comparability of the groups and the appropriateness of the statistical models must be made before the comparisons will be valid. Establishing the validity of these assumptions may be difficult.

Randomized and non-randomized trials of the use of anticoagulant therapy in patients with acute myocardial infarctions have been reviewed by Chalmers et al and the conclusions compared.[25] Of 32 studies, 18 used historical controls and involved a total of 900 patients, eight used non-randomized concurrent controls and involved over 3000 patients, and six were randomized trials with a total of over 3800 patients. The authors reported that 15 of the 18 historical control trials and five of the eight non-randomized concurrent control trials showed statistically significant results favoring the anticoagulation therapy. Only one of the six randomized control trials showed significant results in support of this therapy. Pooling the results of these six randomized trials yielded a statistically significant 20% reduction in total mortality, confirming the findings of the non-randomized studies. Pooling the results of the non-randomized control studies showed a reduction of about 50% in total mortality in the intervention groups, more than twice the decrease seen in the randomized trials. Peto[26] has assumed that this difference in reduction is due to bias. He suggests that since the presumed bias in the non-randomized trials was of the same order of magnitude as the presumed true effect, the non-randomized trials could have yielded positive answers even if the therapy had been of no benefit. Of course, pooling results of several studies can be hazardous. As pointed out by Goldman and Feinstein,[27] not all randomized trials of anticoagulants study the same kind of subjects, use precisely the same intervention or measure the same response variables. And, of course, not all randomized trials are done equally well.

Grace, Muench and Chalmers[28] reviewed studies involving portacaval shunt operations for patients with portal hypertension from cirrhosis. In their review, 34 of 47 non-randomized studies strongly supported the shunt procedure, while only one of the four randomized control trials indicated support for the operation. The authors concluded that the operation should not be endorsed.

Sacks and coworkers have expanded the work by Chalmers et al referenced above, to five other interventions.[29] They concluded that selection biases led historical control studies to inappropriately favor the new interventions. It was also noted that many randomized control trials were of inadequate size, and therefore may have failed to find benefits that truly existed.[30]

The most frequent objections to the use of the randomized control clinical trial are stated by Ingelfinger[31] to be "emotional and ethical." Many clinicians feel that they must not deprive a subject from receiving a new therapy or intervention which they, or someone else, believe to be beneficial, regardless of the validity of the evidence for that claim. The argument aimed at randomization is that in the typical trial it deprives about one-half the subjects from receiving the new and presumed better intervention.

The ethical aspects of randomization have been discussed by several authors.[24,32,33] What is meant by ethical behavior? Most would agree that it means treating the subject with the intervention believed to be best. Of course, what is ethical behavior for one well-informed investigator might not be ethical for another. Presumably, the reason that a clinical trial is being considered at all is that there is uncertainty about the potential benefits of a new intervention. If an investigator believes — for whatever reason — that the new intervention is more beneficial or harmful than the old, he should not participate in the trial. If, on the other hand, he has sufficient doubt about which intervention is better, then he is ethically justified in participating in a randomized clinical trial to settle the question. Shaw and Chalmers[33] argue that under these circumstances, randomization is a more ethical way of practicing medicine than the routine prescribing of medication or therapy that has never been proven to be beneficial by standard scientific methods and could possibly be harmful.

An interesting example of ethics and randomized clinical trials is presented by Silverman[34] and Meier.[35] Administration of high concentrations of oxygen to premature babies was at one time routinely practiced to prevent brain damage due to oxygen deficiency. The suspicion that this procedure might be a cause of retrolental fibroplasia and subsequent blindness led to a clinical trial in which premature babies were randomized to either the accepted high or a lower concentration of oxygen. Results of the trial indicated that the practice of administering high concentrations of oxygen caused blindness.[36] What was once considered to be unethical behavior in withholding high oxygen concentration was later felt to be ethically correct. However, because of possible harm to other organs caused by withholding high concentrations of oxygen, questions regarding the overall benefit or harm remain to be answered.[34]

The lesson from the retrolental fibroplasia study was helpful in justifying a clinical trial involving respiratory distress syndrome in premature babies.[37] This study involved randomization of mothers in premature labor to a corticosteroid or placebo therapy. The assumption was that the corticosteroid would hasten the lung maturation process and protect against respiratory distress syndrome. Pediatricians and obstetricians recalled the oxygen experience and felt ethically able to participate in such a randomized clinical trial before recommending steroid therapy for general use.

Not all clinical studies can use randomized controls. Occasionally, the prevalence of the disease is so rare that a large enough population can not be readily obtained. In such an instance, only case-control studies might be possible. Such studies, which are not clinical trials according to the definition in this book, are discussed in standard epidemiology textbooks.[15,16]

Zelen has proposed a modification of the standard randomized control study.[38] He argued that investigators are often reluctant to recruit prospective trial subjects not knowing to which group the subject will be assigned. Expressing ignorance of optimal therapy compromises the traditional doctor-patient relationship. Zelen, therefore, suggested randomizing eligible subjects before informing them about the trial. Only those assigned to active intervention would be asked if they wish to participate. The control subjects would simply be followed and their outcome monitored. Obviously, such a design could not be blinded. Other major criticisms of this controversial design center around the ethical concerns of not informing subjects that they are enrolled in a trial. The efficiency of the design has also been evaluated.[39] It depends on the proportion of subjects consenting to comply with the assigned intervention. To compensate for this possible inefficiency, one needs to increase the sample size (Chapter 7). The Zelen approach has been tried with varying degrees of success.[40] At the present time, there is insufficient experience to fully evaluate it.

NON-RANDOMIZED CONCURRENT CONTROL STUDIES

Controls in this type of study are subjects treated without the new intervention at approximately the same time as the intervention group is treated. Subjects are allocated to one of the two groups, but by definition this is not a random process. An example of a non-randomized concurrent control study would be a comparison of survival results of patients treated at two institutions, one institution using a new surgical procedure and the other using more traditional medical care.

To some investigators, the non-randomized concurrent control design has advantages over the randomized control design. Those who object to the idea of ceding to chance the responsibility for selecting a person's treatment may favor this design. It is also difficult for some investigators to convince potential subjects of the need for randomization. They find it easier to select a group of people to receive the intervention and would prefer to select the control group by means of matching key characteristics.

The major weakness of the non-randomized concurrent control study is the potential that the intervention group and control group are not strictly comparable. It is difficult to prove comparability because the investigator must assume that he has information on all the important prognostic factors. Selecting a control group by matching on more than a few factors is impractical and the comparability of a variety of other characteristics would still need to be evaluated. In small studies, an investigator is unlikely to find real differences which may exist between groups before the initiation of intervention since there is poor sensitivity to detect such differences. Even for large studies which could detect most differences of real clinical importance, the uncertainty about the unknown or unmeasured factors is still of concern.

Is there, for example, some unknown and unmeasurable process that results in one type of subject's being recruited into one group and not into the other? If all subjects come from one institution, physicians may select subjects into one group based on subtle and intangible factors. In addition, there exists the possibility for subconscious bias in the allocation of subjects to either the intervention or control group. One group might come from a different socioeconomic class than the other group. All of these uncertainties may decrease the credibility of the concurrent but non-randomized control study. For any particular question, the advantages of reduced cost, relative simplicity and investigator and subject acceptance must be carefully weighed against the potential biases before a decision is made to use a non-randomized concurrent control study.

HISTORICAL CONTROLS

In historical control studies, a new intervention is used in a series of subjects and the results are compared to the outcome in a previous series of comparable subjects. Historical controls are, by this definition, non-randomized and non-concurrent.

The argument for using a historical control design is that all new subjects can receive the new intervention. As argued by Gehan and Freireich,[22] many clinicians believe that no subject should be deprived of the possibility of receiving a new therapy or intervention. Some require less supportive evidence than others to accept a new intervention as being beneficial. If an investigator is already of the opinion that the new intervention is beneficial, then he would most likely consider any restriction on its use unethical. Therefore, he would favor a historical control study. In addition, subjects may be more willing to participate in a study if they can be assured of receiving a particular therapy or intervention. Finally, since all new subjects will be on the new intervention and compared to a previous group, the time required to get a given number of subjects on the new intervention will be cut approximately in half. This allows investigators to obtain results faster or do more studies with given resources.

Gehan has emphasized the ethical advantages of historical control studies and pointed out that they have contributed to medical knowledge.[41] Lasagna has argued that medical practitioners have traditionally relied on historical controls when making therapeutic judgments. He maintains that, while sometimes faulty, these judgments are often correct and useful.[42]

Typically, historical control data can be obtained from two sources. First, control group data may be available in the literature. These data are often undesirable because it is difficult, and perhaps impossible, to establish whether the control and intervention groups are comparable in key characteristics at the onset. Even if such characteristics were measured in the same way, the information may not be published and for all practical purposes it will be lost. Second, data may not have been published but may be available on computer files or in medical charts. Such data on control subjects, for example, might be found in a large center which has several ongoing clinical investigations. When one study is finished, the subjects in that study may be used as a control group for some future study. Centers which do successive studies, as in cancer research, will usually have a system for storing and retrieving the data from past studies for use at some future time.

Despite the time and cost benefits, as well as the ethical considerations, historical control studies have potential limitations which should be kept in mind. They are particularly vulnerable to bias. Moertel cites a number of examples of treatments for cancer which have been claimed, on the basis of historical control studies, to be beneficial. Many treatments were declared breakthroughs on the basis of control data as old as 30 years.[43] Pocock[44] identified 19 instances of the same intervention's having been used in two consecutive trials employing similar subjects at the same institution. Theoretically, the mortality in the two groups using the same treatment should be similar. Pocock noted that the difference in mortality rates between such groups ranged from -46% to $+24\%$. Four of the 19 comparisons showed differences significant at the 5% level.

An improvement in outcome for a given disease may be attributed to a new intervention when, in fact, the improvement may stem from a change in the patient population or patient management. Shifts in patient population can be subtle and perhaps undetectable. In a Veterans Administration Urological Research Group study of prostate cancer,[24,45] 2313 patients were randomized to placebo and estrogen treatment groups over a seven-year period. During the last two to three years of the study, no differences were found between the placebo and estrogen groups. However, placebo patients entering in the first two to three years had a shorter survival time than estrogen patients entering in the last two to three years of the study. The reason for the early apparent difference is probably that the patients randomized earlier were older than the later group and thus were at higher risk of death during the period of observation. The results would have been misleading had this been a historical control study and had a concurrent randomized comparison group not been available.

On a broader scale, for reasons which are not entirely clear[46] coronary heart disease has been on the decline in the general population for the past decade. Therefore, any clinical trial in the cardiovascular area involving long-term therapy using historical controls would need to separate the treatment effect from the recent trend, an almost impossible task.

The method by which subjects are selected for a particular study can have a large impact on their comparability with earlier subject groups or general population statistics. In the Coronary Drug Project,[47] an annual total mortality rate of 6% was anticipated in the control group based on rates from a fairly unselected group of myocardial infarction patients. In fact, a control group mortality rate of about 4% was observed, and no significant differences were seen between the intervention groups and the control group. Using the historical control approach, a 33% reduction in mortality might have been claimed for the treatments. One explanation for the discrepancy between anticipated and observed mortality is that entry criteria excluded the most seriously ill subjects.

Shifts in diagnostic criteria for a given disease due to improved technology can cause major changes in the recorded frequency of the disease and in the perceived prognosis of the subjects with the disease. International coding systems and names of diseases change periodically and, unless one is aware of the modifications, prevalence of certain conditions can appear to change abruptly. For example, when the Eighth Revision of the International Classification of Diseases came out in 1968, almost 15% more deaths were assigned to ischemic heart disease than had been assigned in the Seventh Revision.[48]

The characteristics of the population identified as having hypertension today may be quite different than those of a population identified 30 years ago. Today, through education, publicity campaigns and screening programs, people are encouraged to have their blood pressure checked. Many people may be asymptomatic and yet be noted as having hypertension. Several years ago, it is likely that only people with symptoms would have chosen to see a physician. As a result, those classed as hypertensive in the past might have been at different risk than those currently labeled as hypertensive.

A common concern about historical control designs is the accuracy and completeness with which control group data are collected. With the possible exception of special centers which have many ongoing studies, data are generally collected in a non-uniform manner by numerous people with diverse interests in the information. Lack of uniform collection methods can easily lead to incomplete and erroneous records. Data on some important prognostic factors may not have been collected at all. Because of the limitations of data collected historically from medical charts, records from a center which conducts several studies and has a computerized data management system may provide the most reliable historical control data.

Despite the limitations of the historical control study, it does have a place in scientific investigation. As a rapid, relatively inexpensive method of obtaining initial impressions regarding a new therapy, such studies can be important. This is particularly so if investigators understand the potential biases and are willing to miss effective new therapies if bias works in the wrong direction. Bailar et al[49] have identified several features which can strengthen the conclusions to be drawn from historical control studies. These include an *a priori* identification of a reasonable hypothesis and advance planning for analysis.

In some special cases where the diagnosis of a disease is clearly established and the prognosis is well known or the disease highly fatal, a historical control study may be the only reasonable design. The results of penicillin in treatment of pneumococcal pneumonia were so dramatic in contrast to previous experience that no further evidence was really required. Similarly, the benefits of treatment of malignant hypertension became readily apparent from comparisons with previous, untreated populations.[50-52]

The use of prospective registries to characterize patients and evaluate effects of therapy has been advocated.[53,54] Supporters say that a systematic approach to data collection and follow-up can provide information about the local patient population, and can aid in clinical decision making. They argue that clinical trial populations may not be representative of the patients actually seen by a physician. Moon et al have described the use of data bases derived from clinical trials to evaluate therapy.[55] They stress that the high quality data obtained through these sources can reduce the problems of the typical historical control study.

Others[56,57] have emphasized limitations of registry studies such as potential bias in treatment assignment, multiple comparisons, and missing data. Another weakness of prospective data-base registries is that they rely heavily on the validity of the model employed to analyze the data.[58]

CROSS-OVER DESIGNS

The cross-over design is a special case of a randomized control trial and has some appeal to medical researchers. The cross-over design allows each subject to serve as his own control. In the simplest case, namely the two period cross-over design, each subject will receive either intervention or control (A or B) in the first period and the alternative in the succeeding period. The order in which A and B are given to each subject is randomized. Thus, approximately half of the subjects receive the intervention in the sequence AB and the other half in the sequence BA. This is so that any trend from first-period to second-period can be eliminated in the estimate of differences in response.

The advantages and disadvantages of the cross-over design have been described by Brown.[19,21] The appeal of the cross-over design to investigators is that it allows assessment of whether each subject does better on A or B. Since each subject is used twice, once on A and once on B, variability is reduced because the measured effect of the intervention is the difference in an individual subject's response to intervention and control. This reduction in variability enables investigators to use smaller sample sizes to detect a specific difference in response between intervention and control.

In order to use the cross-over design, however, a fairly strict assumption must be made. The assumption is that the effects of the intervention during the first period must not carry over into the second period. This assumption should be independent of which intervention was assigned during the first period and of the subject response. In many clinical trials, such an assumption is clearly inappropriate. If, for example, the intervention during the first period cures the disease, then the subject obviously cannot return to the initial state. In other clinical trials, the cross-over design appears more reasonable. If a drug's effect is to lower blood pressure or heart rate, then a drug-versus-placebo cross-over design might be considered if the drug has no carryover effect once the subject is taken off medication. Obviously, a fatal event cannot serve as the primary response variable in a cross-over trial.

Although the statistical method for checking the assumption of no period-treatment interaction has been described by Grizzle,[59] the test is not so powerful as one would like. What decreases the power of the test is that the mean response of the *AB* group is compared to the mean response of the *BA* group. However, subject variability is introduced in this comparison, which inflates the error term in the statistical test. Thus, the ability to test the assumption of no period-intervention interaction is not sensitive enough to detect important violations of the assumption unless many subjects are used. The basic appeal of this design is to avoid between-subject variation in estimating the intervention effect, thereby requiring a smaller sample size. Yet the ability to justify the use of the design still depends on a test for carryover that includes between-subject variability. This may weaken the main rationale for the cross-over design. Because of this insensitivity, the cross-over design is not as attractive as it at first appears. Brown[19,21] and Hills and Armitage[60] discourage the use of the cross-over design in general. Only if there is substantial evidence that the therapy has no carryover effects, and the scientific community is convinced by that evidence, should a cross-over design be used.

WITHDRAWAL STUDIES

A number of studies have been conducted in which the subjects on a particular treatment for a chronic disease are taken off therapy or have the dosage reduced. The objective is to assess response to the discontinuation or reduction. This design may be validly used to evaluate the duration of benefit of an intervention already known to be useful. For example, subsequent to the Hypertension Detection and Follow-up Program,[61] which demonstrated the benefits of treating mild and moderate hypertension, several investigators withdrew a sample of subjects with controlled blood pressure from antihypertensive therapy.[62] One aim was to see if life-long therapy was necessary.

Withdrawal studies have also been used to assess the efficacy of an intervention that has never conclusively been shown to be beneficial. An example is the 60 Plus Reinfarction Study.[63] Subjects doing well on oral anticoagulant therapy since their myocardial infarction, an average of six years earlier, were randomly continued on anticoagulants or assigned to placebo. Those who stayed on the intervention had lower mortality (not statistically significant) and a clear reduction in nonfatal reinfarction.

One serious limitation of this type of study is that a highly selected sample is evaluated. Only those subjects who physicians thought were benefiting from the intervention were likely to have been on it for several years. Anyone who had major adverse effects from the drug would have been taken off and, therefore, not been eligible for the withdrawal study. Thus, this design can overestimate benefit and underestimate toxicity. Another drawback is that both subjects and disease states change over time. In the above example, it cannot validly be claimed that anticoagulants were shown to be beneficial if administered immediately after an infarction.

If withdrawal studies are conducted, the same standards should be adhered to that are used with other designs. Randomization, blinding where feasible, unbiased assessment, and proper data analysis are as important here as in other settings.

FACTORIAL DESIGN

In the simple case, the factorial design attempts to evaluate two interventions compared to control in a single experiment (Table 4-1).[2-4] Given the cost and effort in recruiting subjects and conducting clinical trials, getting two experiments done at once is appealing. A factorial design was carried out in the Canadian transient ischemic attack study where aspirin and sulfinpyrazone were compared with placebo.[64] The study indicated that aspirin was useful in reducing the frequency of stroke, that sulfinpyrazone had no effect, and that the combination of drugs was no better than aspirin alone.

Table 4-1
Factorial Design

	Intervention X	Control
Intervention Y	a	b
Control	c	d

cell	intervention
a	X + Y
b	Y + control
c	X + control
d	control + control

The appeal of this design might suggest that there really is a "free lunch." However, every design has strengths and weaknesses. A concern with the factorial design is the possibility of the existence of interaction and its impact on the sample size. Interaction means that the effect of intervention X differs depending upon the presence or absence of intervention Y, or vice versa. It is more likely to occur when the two drugs are expected to have related mechanisms of action.

If one could safely assume there were no interactions, one can show that with a modest increase in sample size two experiments can be conducted in one; one which is considerably smaller than the sum of two independent trials under the same design specifications. However, if one cannot reasonably rule out interaction, one should statistically test for its presence. As is true for the cross-over design, the power for testing for interaction is less than the power for testing for the main effects of interventions (cells a + c versus b + d or cells a + b versus c + d). Thus, to obtain satisfactory power to detect interaction, the total sample size must be increased. The extent of the increase depends on the degree of interaction, which may not be known until the end of the trial. The larger the interaction, the smaller the increase in sample size. If an interaction is detected, or perhaps only suggested, the comparison of intervention X would have to be done individually for intervention Y and its control (cell a versus b and cell c versus d). The power for these comparisons is obviously reduced.

The factorial design has some distinct advantages. If the interaction of two interventions is important to determine, or if there is little chance of interaction, then such a design with appropriate sample size can be very informative and efficient. However, the added complexity, impact on recruitment and compliance, and potential adverse effects of "polypharmacy" must be considered.

GROUP ALLOCATION DESIGNS

In group allocation or composite randomization designs, a group of individuals, a clinic or a community are randomized to a particular intervention or control.[65-68] These designs have been proposed in cancer trials where a clinic or physician may have difficulty approaching people about the idea of randomization. Giving all subjects a specific intervention, however, may be quite acceptable. In this design, the basic sampling units are groups, not subjects. This means that the effective sample does not consist of the total number of subjects, and the design is not as efficient as the traditional one. If the response rates vary across clinics or groups, efficiency is further decreased.

HYBRID DESIGNS

Pocock[69] has argued that if a substantial amount of data is available from historical controls, then a hybrid, or combination design could be considered. Rather than a 50/50 allocation of subjects, a smaller proportion could be randomized to control, permitting most to be assigned to the new intervention. A number of criteria must be met in order to combine the historical and randomized controls. These include the same entry criteria and evaluation factors, and subject recruitment by the same clinic or investigator. The data from the historical control subjects must also be fairly recent. This approach requires fewer subjects to be entered into a trial. Machin, however, cautions that if biases introduced from the non-randomized subjects (historical controls) are substantial, more subjects might have to be randomized to compensate than would be the case in a corresponding fully randomized trial.[70]

STUDIES OF EQUIVALENCY

In studies of equivalency, or trials with "positive controls," the objective is to test whether a new intervention is as good as an established one. Sample size issues for this kind of trial are discussed in Chapter 7. Several design aspects also need to be considered. The control or standard treatment must have been shown to be effective; that is, truly better than placebo or no therapy. The circumstances under which the control was found to be useful ought to be reasonably close to those of the planned trial. Similarity of populations, concomitant therapy, and dosage are important. These requirements also mean that the trials which demonstrated efficacy of the standard should be recent and properly designed and conducted.

As emphasized in Chapter 7, the investigator must specify what he means by equivalence. It cannot be statistically shown that two therapies are identical, as an infinite sample size would be required. Therefore, if the intervention falls sufficiently close to the standard, as defined by reasonable boundaries, the two are claimed to be the same. In this situation, outcomes other than the primary response variable become important factors. A judgment regarding use of similar treatments can depend on frequency and severity of adverse effects and changes in quality of life. Thus, the study must be designed to allow for proper evaluation of these factors (Chapters 11 and 12).

REFERENCES

1. Fisher, R.A. *Statistical Methods for Research Workers*. Edinburgh: Oliver and Boyd, 1925.
2. Fisher, R.A. *The Design of Experiments*. Edinburgh: Oliver and Boyd, 1935.
3. Cochran, W.G., and Cox, G.M. *Experimental Designs*. 2nd Ed. New York: John Wiley and Sons, 1957.

48

4. Cox, D.R. *Planning of Experiments*. New York: John Wiley and Sons, 1958.

5. Bull, J.P. The historical development of clinical therapeutic trials. *J Chronic Dis*. 10:218–248, 1959.

6. Eliot, M.M. The control of rickets: preliminary discussion of the demonstration in New Haven. *JAMA* 85:656–663, 1925.

7. Hill, A.B. Observation and experiment. *N Engl J Med*. 248:995–1001, 1953.

8. Macfarlane, G. *Howard Florey: The Making of a Great Scientist*. Oxford: Oxford University Press, 1979:11–12.

9. Gocke, D.J. Fulminant hepatitis treated with serum containing antibody to Australia antigen. *N Engl J Med*. 284:919, 1971.

10. Acute Hepatic Failure Study Group. Failure of specific immunotherapy in fulminant type B hepatitis. *Ann Intern Med*. 86:272–277, 1977.

11. Snow, J.B., Jr., and Kimmelman, C.P. Assessment of surgical procedures for Ménière's disease. *Laryngoscope* 89:737–747, 1979.

12. Armitage, P. *Statistical Methods in Medical Research*. New York: John Wiley and Sons, 1971.

13. Brown, B.W., and Hollander, M. *Statistics: A Biomedical Introduction*. New York: John Wiley and Sons, 1977.

14. Feinstein, A.R. *Clinical Biostatistics*. St. Louis: The C.V. Mosby Company, 1977.

15. MacMahon, B., and Pugh, T.F. *Epidemiology: Principles and Methods*. Boston: Little, Brown, 1970.

16. Lilienfeld, A.M. *Foundations of Epidemiology*. New York: Oxford University Press, 1976.

17. Srivastava, J.N. (Editor) *A Survey of Statistical Design and Linear Models*. Amsterdam: North-Holland, 1975.

18. Peto, R., Pike, M.C., Armitage, P., et al. Design and analysis of randomized clinical trials requiring prolonged observation of each patient. I. Introduction and design. *Br J Cancer*. 34:585–612, 1976.

19. Brown, B.W., Jr. Statistical controversies in the design of clinical trials—some personal views. *Controlled Clin Trials* 1:13–27, 1980.

20. Pocock, S.J. Allocation of patients to treatment in clinical trials. *Biometrics* 35:183–197, 1979.

21. Brown, B.W. The crossover experiment for clinical trials. *Biometrics* 36:69–80, 1980.

22. Gehan, E.A., and Freireich, E.J. Non-randomized controls in cancer clinical trials. *N Engl J Med*. 290:198–203, 1974.

23. Weinstein, M.C. Allocation of subjects in medical experiments. *N Engl J Med*. 291:1278–1285, 1974.

24. Byar, D.P., Simon, R.M., Friedewald, W.T. et al. Randomized clinical trials: perspectives on some recent ideas. *N Engl J Med*. 295:74–80, 1976.

25. Chalmers, T.C., Matta, R.J., Smith, H., and Kunzler, A.M. Evidence favoring the use of anticoagulants in the hospital phase of acute myocardial infarction. *N Engl J Med*. 297:1091–1096, 1977.

26. Peto, R. Clinical trial methodology. *Biomedicine* (Special issue) 28:24–36, 1978.

27. Goldman, L., and Feinstein, A.R. Anticoagulants and myocardial infarction: the problems of pooling, drowning, and floating. *Ann Intern Med*. 90:92–94, 1979.

28. Grace, N.D., Muench, H., and Chalmers, T.C. The present status of shunts for portal hypertension in cirrhosis. *Gastroenterology* 50:684–691, 1966.

29. Sacks, H., Chalmers, T.C., and Smith, H., Jr. Randomized versus historical controls for clinical trials. *Am J Med*. 72:233–240, 1982.

30. Sacks, H.S., Chalmers, T.C., and Smith, H., Jr. Sensitivity and specificity of clinical trials: randomized v historical controls. *Arch Intern Med*. 143:753–755, 1983.

31. Ingelfinger, F.J. The randomized clinical trial (editorial). *N Engl J Med*. 287:100–101, 1972.

32. Chalmers, T.C., Black, J.B., and Lee, S. Controlled studies in clinical cancer research. *N Engl J Med*. 287:75–78, 1972.

33. Shaw, L.W., and Chalmers, T.C. Ethics in cooperative clinical trials. *Ann NY Acad Sci*. 169:487–495, 1970.

34. Silverman, W.A. The lesson of retrolental fibroplasia. *Sci Am*. 236:100–107, June 1977.

35. Meier, P. Terminating a trial - the ethical problem. *Clin Pharmacol Ther*. 25:633–640, 1979.

36. Lanman, J.T., Guy, L.P., and Dancis, J. Retrolental fibroplasia and oxygen therapy. *JAMA* 155:223–226, 1954.

37. Collaborative Group on Antenatal Steroid Therapy. Effect of antenatal dexamethasone administration on the prevention of respiratory distress syndrome. *Am J Obstet Gynecol*. 141:276–287, 1981.

38. Zelen, M. A new design for randomized clinical trials. *N Engl J Med*. 300:1242-1245, 1979.

39. Anbar, D. The relative efficiency of Zelen's prerandomization design for clinical trials. *Biometrics* 39:711–718, 1983.

40. Ellenberg, S.S. Randomization designs in comparative clinical trials. *N Engl J Med*. 310:1404–1408, 1984.

41. Gehan, E.A. The evaluation of therapies: historical control studies. *Stat Med*. 3:315–324, 1984.

42. Lasagna, L. Historical controls: the practitioner's clinical trials. *N Engl J Med*. 307:1339–1340, 1982.

43. Moertel, C.G. Improving the efficiency of clinical trials: a medical perspective. *Stat Med*. 3:455–465, 1984.

44. Pocock, S.J. Letter to the editor. *Br Med J*. 1:1661, 1977.

45. Veterans Administration Co-operative Urological Research Group. Treatment and survival of patients with cancer of the prostate. *Surg Gynecol Obstet*. 124:1011–1017, 1967.

46. *Proceedings of the Conference on the Decline in Coronary Heart Disease Mortality*. Edited by R.J. Havlik and M. Feinleib. Washington, D.C.: NIH Publication No. 79-1610, 1979.

47. Coronary Drug Project Research Group. Clofibrate and niacin in coronary heart disease. *JAMA* 231:360–381, 1975.

48. Rosenberg, H.M., and Klebba, A.J. Trends in cardiovascular mortality with a focus on ischemic heart disease: United States, 1950–1976. Edited by R. Havlik and M. Feinleib. In *Proceedings of the Conference on the Decline in Coronary Heart Disease Mortality*. Washington, D.C.: NIH Publication No. 79-1610, 1979.

49. Bailar, J.C. III, Louis, T.A., Lavori, P.W., et al. Studies without internal controls. *N Engl J Med*. 311:156–162, 1984.

50. Dustan, H.P. Schneckloth, R.E., Corcoran, A.C. and Page, I.H. The effectiveness of long-term treatment of malignant hypertension. *Circulation* 18: 644–651, 1958.

51. Bjork, S., Sannerstedt, R., Angervall, G., and Hood, B. Treatment and

50

prognosis in malignant hypertension: clinical follow-up study of 93 patients on modern medical treatment. *Acta Med Scand.* 166:175–187, 1960.

52. Bjork, S., Sannerstedt, R., Falkheden, T., and Hood, B. The effect of active drug treatment in severe hypertensive disease: an analysis of survival rates in 381 cases on combined treatment with various hypotensive agents. *Acta Med Scand.* 169:673–689, 1961.

53. Starmer, C.F., Lee, K.L., Harrell, F.E., et al. On the complexity of investigating chronic illness. *Biometrics* 36:333–335, 1980.

54. Hlatky, M.A., Lee, K.L., Harrell, F.E., Jr., et al. Tying clinical research to patient care by use of an observational database. *Stat Med.* 3:375–384, 1984.

55. Moon, T.E., Jones, S.E., Bonadonna, G., et al. Using a data base of protocol studies to evaluate therapy: a breast cancer example. *Stat Med.* (in press).

56. Byar, D.P. Why databases should not replace randomized clinical trials. *Biometrics* 36:337–342, 1980.

57. Dambrosia, J.M., and Ellenberg, J.H. Statistical considerations for a medical database. *Biometrics* 36:323–332, 1980.

58. Mantel, N. Cautions on the use of medical databases. *Stat Med.* 2:355–362, 1983.

59. Grizzle, J.E. The two period change-over design and its use in clinical trials. *Biometrics* 21:467–480, 1965.

60. Hills, M., and Armitage, P. The two-period cross-over clinical trial. *Br J Clin Pharmacol.* 8:7–20, 1979.

61. Hypertension Detection and Follow-up Program Cooperative Group. Five-Year findings of the Hypertension Detection and Follow-Up Program. 1. Reduction in mortality of persons with high blood pressure, including mild hypertension. *JAMA* 242:2562–2571, 1979.

62. Stamler, R., Berman, R., Stamler, J., et al. Nonpharmacologic control of hypertension (abstract). *Circulation* 66(Suppl II):II-328, 1982.

63. Report of the Sixty Plus Reinfarction Study Research Group. A double-blind trial to assess long-term oral anticoagulant therapy in elderly patients after myocardial infarction. *Lancet* ii:989–994, 1980.

64. The Canadian Cooperative Study Group. A randomized trial of aspirin and sulfinpyrazone in threatened stroke. *N Engl J Med.* 299:53–59, 1978.

65. Donner, A., Birkett, N., and Buck, C. Randomization by cluster: sample size requirements and analysis. *Am J Epidemiology* 114:906–914, 1981.

66. Armitage, P. The role of randomization in clinical trials. *Stat Med.* 1:345–352, 1982.

67. Simon, R. Composite randomization designs for clinical trials. *Biometrics* 37:723–731, 1981.

68. Cornfield, J. Randomization by group: a formal analysis. *Am J Epidemiology* 108:100–102, 1978.

69. Pocock, S.J. The combination of randomized and historical controls in clinical trials. *J Chron Dis.* 29:175–188, 1976.

70. Machin, D. On the possibility of incorporating patients from non-randomising centres into a randomised clinical trial. *J Chron Dis.* 32:347–353, 1979.

CHAPTER 5

The Randomization Process

The randomized control clinical trial is the standard by which all trials are judged since other designs have certain undesirable features. In the simplest case, randomization is a process by which each subject has the same chance of being assigned to either intervention or control. An example would be the toss of a coin, in which heads indicates intervention group and tails indicates control group. Even in the more complex randomization strategies, the element of chance underlies the allocation process. Of course, neither subject nor investigator should know what the assignment will be before the subject's decision to enter the study. Otherwise, the benefits of randomization can be lost. The role that randomization plays has been discussed in Chapter 4 as well as by Armitage[1] and Byar et al.[2]

FUNDAMENTAL POINT

Randomization tends to produce study groups comparable with respect to known as well as unknown risk factors, removes investigator bias in the allocation of subjects, and guarantees that statistical tests will have valid significance levels.

Several methods for randomly allocating subjects are currently in use. This chapter will present these methods and consider the advantages and disadvantages of each. Unless stated otherwise, it can be assumed that the randomization strategy will allocate subjects into two groups, an intervention group and a control group. However, many of the methods described here can easily be generalized for use with more than two groups.

FIXED ALLOCATION RANDOMIZATION

Fixed allocation procedures assign the interventions to subjects with a pre-specified probability, usually equal, and that allocation probability is not altered as the study progresses. Zelen[3] and Pocock[4] discuss a number of methods by which fixed allocation is achieved, and we will review three of these — simple, blocked, and stratified.

Our view is that allocation to intervention and control should be equal. Peto,[5] among others, has suggested an unequal allocation ratio, such as 2:1, of intervention to control. The rationale for a 2:1 allocation is that the study may slightly lose sensitivity but may gain more information about subject responses to the new intervention, such as toxicity and side effects. In some instances, less information may be needed about the control group and, therefore, fewer control subjects are required. If the intervention turns out to be beneficial, more study subjects would benefit than under an equal allocation scheme. However, new interventions may also turn out to be harmful, in which case more subjects would receive them under the unequal allocation strategy. Although the loss of sensitivity or power may be less than 5% for allocation ratios approximately between ½ and ⅔,[6,7] more often than not, studies should have the most powerful design possible. We also believe that equal allocation is more consistent with the view of indifference toward which of the two groups a subject is assigned. Unequal allocation may indicate to the subject and to his personal physician that one intervention is preferred over the other. Thus, equal allocation will be presumed throughout the following discussion unless otherwise indicated.

Simple Randomization

The most elementary form of randomization is best illustrated by a few examples. One simple method is to toss an unbiased coin each time a subject is eligible to be randomized. If the coin turns up heads, the subject is assigned to group A; if tails, to group B. Using this procedure, approximately one half of the subjects will be in group A and one half in group B. Instead of tossing a coin to generate a randomization schedule, a random digit table on which the equally likely digits 0 to 9 are arranged by rows and columns is usually used to accomplish simple randomization. By randomly selecting a certain row (column) and observing the sequence of digits in that row (column) A could be assigned, for example, to those subjects for whom the next digit was even and B to those for whom the next digit was odd. This process produces a sequence of assignments which is random in order, and each subject has an equal chance of being assigned to A or B.

For large studies, a more convenient method for producing a randomization schedule is to use a random number producing algorithm, available on most digital computer systems. A simple randomization procedure might assign subjects to group A with probability P and subjects to group B with probability 1-P. One computerized process for simple randomization is to use a uniform random number algorithm to produce random numbers in the interval from 0.0 to 0.999. Using a uniform random number generator, a random number can be produced for each subject. If the random number is between 0 and P, the subject would be assigned to group A; otherwise to group B. If equal allocation between A and B is not desired ($P \neq \frac{1}{2}$), then P can be set to the desired proportion in the algorithm and the study will have, on the average, a proportion P of the subjects in group A.

This procedure can be adapted easily to more than two groups. Suppose, for example, the trial has three groups, A, B and C, and subjects are to be randomized such that a subject has a ¼ chance of being in group A, a ¼ chance of being in group B, and a ½ chance of being in group C. By dividing the interval 0 to 1 into three pieces of length ¼, ¼, and ½, random numbers generated will have probabilities of ¼, ¼ and ½, respectively, of falling into each subinterval. Specifically, the intervals would be 0–0.249, 0.25–0.499, 0.50–0.999. Then any subject whose random number falls between 0 and 0.249 is assigned A, any subject whose random number falls between 0.25 and 0.499 is assigned B and the others, C. For equal allocation, the interval would be divided into thirds and assignments made accordingly.

The advantage of this simple randomization procedure is that it is easy to implement. The major disadvantage is that, although in the long run the number of subjects in each group will be in the proportion anticipated, at any point in the randomization, including the end, there could be a substantial imbalance. This is true particularly if the sample size is small. For example, if 20 subjects are randomized with equal probability to two treatment groups, the chance of a 12:8 split or worse is approximately 50%. For 100 subjects, the chance of the same ratio (60:40 split) or worse is only 5%. While such imbalances do not cause the statistical tests to be invalid, they do reduce ability to detect true differences between the two groups. In addition, such imbalances appear awkward and may lead to some loss of credibility for the trial, especially for the person not oriented to statistics.

Some investigators incorrectly believe that an alternating assignment of subjects to the intervention and the control groups (eg, $ABABAB...$) is a form of randomization. However, no random component exists in this type of allocation except perhaps for the first subject. A major criticism of this method is that, in a single-blind or unblinded study, the investigators know the next assignment, which could lead to a bias in the selection of subjects. Even in a double-blind study, if the blind is broken on one subject, as sometimes happens, the entire sequence of assignments is known. Therefore, this type of allocation method should be avoided.

Blocked Randomization

Blocked randomization, sometimes called permuted block randomization, was described by Hill[8] in 1951. It is used in order to avoid the imbalance in the number of subjects assigned to each group, an imbalance which could occur in the simple randomization procedure. Blocked randomization guarantees that at no time during randomization will the imbalance be large and that at certain points the number of subjects in each group will be equal.

If subjects are randomly assigned with equal probability to groups A or B, then for each block of even size (for example, 4, 6 or 8) one half of the subjects will be assigned to A and the other half to B. The order in which the interventions are assigned in each block is randomized, and this process is repeated for consecutive blocks of subjects until all subjects are randomized. For example, the investigators may want to ensure that after every fourth randomized subject, the number of subjects in each intervention group is equal. Then a block of size 4 would be used and the process would randomize the order in which two A's and two B's are assigned for every consecutive group of four subjects entering the trial. One may write down all the ways of arranging the groups and then randomize the order in which these combinations are selected. In the case of blocksize 4, there are six possible combinations of group assignments: $AABB$, $ABAB$, $BAAB$, $BABA$, $BBAA$, and $ABBA$. One of these arrangements is selected at random and the four subjects are assigned accordingly. This process is repeated as many times as needed.

Another method of blocked randomization may also be used. In this method for randomizing the order of assignments within a block of size b, a random number between 0 and 1 for each of the b assignments (half of which are A and the other half B) is obtained. The example below illustrates the procedure.

Example:

Assignment	Random Number
A	0.069
A	0.734
B	0.867
B	0.312

The assignments then are ranked according to the random numbers. This leads to the assignment order of $ABAB$. This process is repeated for another set of four subjects until all subjects have been randomized.

The advantage of blocking is that balance between the number of subjects in each group is guaranteed during the course of randomization. The number in each group will never differ by more than $b/2$ when b is the length of the block. This can be important for at least two reasons. First, if the type of subject recruited for the study changes during the entry period, blocking will produce more comparable groups. For example, an investigator may use sources of potential subjects sequentially. Subjects from different sources may vary in severity of illness or other crucial respects. One source, with the more seriously ill subjects, may be used early during enrollment and another source, with healthier subjects, late in enrollment.[2] If the randomization were not blocked, more of the seriously ill subjects might be randomized to one group. Because the later subjects are not as sick, this early imbalance would not be corrected. A second advantage of blocking is that if the trial should be terminated before enrollment is completed, balance will exist in terms of number of subjects randomized to each group.

A potential problem with blocked randomization is that if the blocking factor b is known by the study staff and the study is not double-blind, the assignment for the last person entered in each block is known before randomization of that person. If the blocking factor is 4 and the first three assignments are ABB, then the next assignment must be A. This could, of course, permit a bias in the selection of every fourth subject to be entered. Clearly, there is no reason to make the blocking factor known. However, in a study that is not double-blind, with a little ingenuity the staff can soon discover the blocking factor. For this reason, repeated blocks of size 2 should not be used. On a few occasions, perhaps as an intellectual challenge, investigators or their clinic staff have attempted to break the randomization scheme. This curiosity is natural but nevertheless can cause problems in the integrity of the randomization process. To avoid this problem in the trial that is not double-blind, the blocking factor can be varied as the recruitment continues. In fact, after each block has been completed, the size of the next block could be determined in a random fashion from a few possibilities such as 2, 4, 6, and 8. The probabilities of selecting a block size can be set at whatever one wishes. For example, the probabilities of selecting block sizes 2, 4, 6, and 8 can be $\frac{1}{6}$, $\frac{1}{6}$, $\frac{1}{3}$, and $\frac{1}{3}$ respectively. Randomly selecting the block size makes it very difficult to determine where blocks start and stop and thus determine the next assignment.

A disadvantage of blocked randomization is that, from a strictly theoretical point of view, analysis of the data is more cumbersome than if simple randomization were used. The data analysis performed at the end of the study should reflect the randomization process actually performed. This requirement would complicate the analysis because many analytical methods assume a simple randomization. In their analysis of the data most investigators ignore the fact that the randomization was blocked. Matts and McHugh[9] studied this problem and concluded that the measurement of variability used in the statistical analysis is not exactly correct if the blocking is ignored. Since blocking guarantees balance between the two groups and, therefore, increases the power of a study, blocked randomization with the appropriate analysis is more powerful than not blocking or blocking and then ignoring it in the analysis. Statisticians recognize the problem and feel that, at worst, they are being conservative by ignoring the fact that the randomization was blocked. That is, the study will have probably slightly less power than it could have with the correct analysis, and the true significance level is less than that computed.

Stratified Randomization

One of the objectives in allocating subjects is to achieve between-group comparability of certain characteristics known as prognostic or risk factors.[3,10] Measured at baseline, these are factors which correlate with subsequent subject response or outcome. Investigators may become concerned when prognostic factors are not evenly distributed between intervention and control groups. As indicated previously, randomization tends to produce groups which are, on the average, similar in their entry characteristics, both known and unknown. This is a concept likely to be true for large studies or for many small studies when averaged. For any single study, especially a small study, there is no guarantee that all baseline characteristics will be similar in the two groups. In the multicenter Aspirin Myocardial Infarction Study which had 4524 subjects, the top 20 cardiovascular prognostic factors for total mortality identified in the Coronary Drug Project[11] were compared in the intervention and control groups and no major differences were found (Furberg, CD, unpublished data). However, individual clinics, with an average of 150 subjects, showed considerable imbalance for many variables between the groups. Imbalances in prognostic factors can be dealt with either after the fact by using stratification in the analysis (Chapter 16) or can be prevented by using stratification in the randomization. Stratified randomization is a method which helps achieve comparability between the study groups for those factors considered.

Stratified randomization requires that the prognostic factors be measured either before or at the time of randomization. If a single factor is used, it is divided into two or more subgroups or strata (eg, age 30 to 34 years, 35 to 39 years, 40 to 44 years). If several factors are used, a stratum is formed by selecting one subgroup from each of them. The total number of strata is the product of the number of subgroups in each factor. The stratified randomization process involves measuring the level of the selected factors for a subject, determining to which stratum he belongs and performing the randomization within that stratum.

Within each stratum, the randomization process itself could be simple randomization, but in practice most clinical trials use some blocked randomization strategy. Under a simple randomization process, imbalances in the number in each group within the stratum could easily happen and thus defeat the purpose of the stratification. Blocked randomization is, as described previously, a special kind of stratification. However, this text will restrict use of the term blocked randomization to stratifying over time, and use stratified randomization to refer to stratifying on factors other than time. Some confusion may arise here because early texts on design used the term blocking as this text uses the term stratifying. However, the definition herein is consistent with current usage in clinical trials.

As an example of stratified randomization with a block size of 4, suppose an investigator wants to stratify on age, sex and smoking history. One possible classification of the factors would be three 10-year age levels and three smoking levels.

Age	Sex	Smoking History
1. 40–49 yrs.	1. Male	1. Current Smoker
2. 50–59 yrs.	2. Female	2. Ex-Smoker
3. 60–69 yrs.		3. Never Smoked

Thus, the design has $3 \times 2 \times 3 = 18$ strata. The randomization for this example appears in Table 5-1.

Table 5-1
Stratified Randomization with Block Size of Four

Strata	Age	Sex	Smoking	Group Assignment
1	40–49	M	Current	*ABBA BABA....*
2	40–49	M	Ex	
3	40–49	M	Never	
4	40–49	F	Current	
5	40–49	F	Ex	
6	40–49	F	Never	
7	50–59	M	Current	
8	50–59	M	Ex	
9	50–59	M	Never	
10	50–59	F	Current	
11	50–59	F	Ex	
12	50–59	F	Never	
	(etc.)			

Subjects who were between 40-49 years old, male and current smokers; that is, in stratum 1, would be assigned to groups *A* or *B* in the sequences *ABBA BABA*.... Similarly, random sequences would appear in the other strata.

Small studies are the ones most likely to require stratified randomization, because in large studies, the magnitude of the numbers increases the chance of comparability of the groups. In the example shown above, with three levels of the first factor (age), two levels of the second factor (sex), and three levels of the third factor (smoking history), 18 strata have been created. As factors are added and the levels within factors are refined, the number of strata increase rapidly. If the example with 18 strata had 100 subjects to be randomized, then only five to six subjects would be expected per stratum if the study population were evenly distributed among the levels. Since the population is most likely not evenly distributed over the strata, some strata would actually get fewer than five to six subjects. If the number of strata were increased, the number of subjects in each stratum would be even fewer. Pocock and Simon[12] showed that increased stratification in small studies can be self-defeating because of the sparseness of data within each stratum. Thus, only important variables should be chosen and the number kept to a minimum.

In addition to making the two study groups appear comparable with regard to specified factors, the power of the study can be increased by taking the stratification into account in the analysis. Stratified randomization, in a sense, breaks the trial down into smaller trials. Subjects in each of the smaller trials belong to the same stratum. This reduces variability in group comparisons if the stratification is used in the analysis. Reduction in variability allows a study of a given size to detect smaller group differences in response variables or to detect a specified difference with fewer subjects.[13,14]

Sometimes the variables initially thought to be most prognostic and, therefore used in the stratified randomization, turn out to be unimportant. Other factors may be identified later which, for the particular study, are of more importance. If randomization is done without stratification, then analysis can take into account those factors of interest and will not be complicated by factors thought to be important at the time of randomization. It has been argued[6] that there usually does not exist a need to stratify at randomization because stratification at the time of analysis will achieve nearly the same expected power. This issue of stratifying pre- versus post-randomization has been widely discussed.[15-18] It appears for a large study that stratification after randomization provides nearly equal efficiency to stratification before randomization. However, for studies of 100 subjects or fewer, stratifying the randomization using two or three prognostic factors may achieve greater power.

Stratified randomization is not the complete solution to all potential problems of baseline imbalance. Another strategy for small studies with many prognostic factors is considered below in the section on adaptive randomization.

In multicenter trials, centers vary with respect to the type of subjects randomized as well as the quality and type of care given to subjects during follow-up. Thus, the center may be an important factor related to subject outcome, and the randomization process should be stratified accordingly.[19] Each center then represents, in a sense, a replication of the trial, though the number of subjects within a center is not adequate to answer the primary question. Nevertheless, results at individual centers can be compared to see if trends are consistent with overall results. Another reason for stratification by center is that if a center should have to leave the study, the balance in prognostic factors in other centers would not be affected.

ADAPTIVE RANDOMIZATION PROCEDURES

The randomization procedure described in the sections on fixed allocation were non-adaptive strategies. In contrast, adaptive procedures change the allocation probabilities as the study progresses. Two types of adaptive procedures will be considered here. First, the section will discuss methods which adjust or adapt the allocation probabilities according to imbalances in numbers of subjects or in baseline characteristics between the two groups. Second, it will briefly review adaptive procedures that adjust allocation probabilities according to the responses of subjects to the assigned intervention.

Baseline Adaptive Randomization Procedures

The *Biased Coin Randomization* procedure, originally discussed by Efron,[20] attempts to balance the number of subjects in each treatment group based on the previous assignments, but does not take subject responses into consideration. The purpose of the algorithm is basically to randomize the allocation of subjects to groups A and B with equal probability as long as the number of subjects in each group is equal. If an imbalance occurs and the difference in numbers is greater than some prespecified value, the allocation probability (P) is adjusted so that the probability is higher for the group with fewer subjects. The investigator can determine the value of the allocation probability he wishes to use. The larger the value of P, the faster the imbalance will be corrected, while the nearer P is to 0.5, the slower the correction. Efron suggests an allocation probability of $P = \frac{2}{3}$ when a correction is indicated. Since much of the time $P \neq \frac{1}{2}$, the process has been named the "biased coin" method. This procedure can be modified to include consideration of the number of consecutive assignments to the same group and the length of such a run.

This approach, from a strictly theoretical point of view, demands a cumbersome data analysis process. The correct analysis requires that the significance level for the test statistic be determined by considering all possible sequences of assignments which could have been made in repeated experiments using the same biased coin allocation rule where no group differences are assumed to exist. Although this is feasible to do with digital computers, analysis is not easy. As with the blocked randomization scheme, the analysis often ignores this requirement. Efron[20] argues that it is probably not necessary to take the biased coin randomization into account in the analysis, especially for larger studies. However, a test statistic which ignores the biased coin randomization will not provide the correct variance term. Most often, the variance will be larger than it would be with proper calculation, thus giving a conservative test in the sense that the probability of rejecting the null hypothesis is less than it would be if the proper analysis were used.

One possible advantage of the biased coin approach over the blocked randomization scheme is that the investigator cannot determine the next assignment by discovering the blocking factor. However, the biased coin method does not appear to be as widely used as the blocked randomization scheme because of its complexity.

Other stratification methods are adaptive in the sense that intervention assignment probabilities for a subject are a function of the distribution of the prognostic factors for subjects already randomized. This concept was suggested by Efron[20] as an extension of the biased coin method and also has been discussed in depth by Pocock and Simon,[12] Freedman and White,[21] Begg and Iglewicz,[22] and Atkinson.[23] In a simple example, if age is a prognostic factor and one study group has more older subjects than the other, the allocation scheme is such that the next several older subjects would most likely be randomized to the group which currently has fewer older subjects. Various methods can be used as the measure of imbalance in prognostic factors. In general, adaptive stratification methods incorporate several prognostic factors in making an "overall assessment" of the group balance or lack of balance. Subjects are then assigned to a group in a manner which will tend to correct an existing imbalance or cause the least imbalance in prognostic factors. This method is sometimes called "minimization" because imbalances in the distribution of prognostic factors are minimized. However, as indicated in the Appendix, the term minimization is also used to refer to a very specific form of adaptive stratification.[24] Generalization of this strategy exists for more than two groups. Development of these methods was motivated in part by the previously described problems with non-adaptive stratified randomization for small studies. Adaptive methods do not have empty or near empty strata because randomization does not take place within a stratum although prognostic factors are used. These methods are being used, especially in clinical trials of cancer where several prognostic factors need to be balanced, and the sample size is typically 100–200 subjects.

The major advantage of this procedure is that it protects against a severe baseline imbalance for important prognostic factors. Overall marginal balance is maintained in the intervention groups with respect to a large number of prognostic factors. One disadvantage is that adaptive stratification is operationally more difficult to carry out, especially if a large number of factors are considered. However, small programmable calculators can minimize most of the computational problems. White and Freedman[25] developed a simplified version of the adaptive stratification method by using a set of specially arranged index cards. Another disadvantage is that the data analysis is complicated, from a strict viewpoint, by the randomization process. The appropriate analysis involves simulating on a computer the assignment of subjects to groups by the actual adaptive strategy used. Replication of the simulation, assuming that no group differences exist, generates the significance level of the statistical test to be used.

Statisticians are not likely to go through the simulation experiments but would rather use the conventional statistical test and standard critical values to determine significance levels. As with other non-simple randomization procedures, this strategy is probably somewhat conservative. The impact of one minimization approach on the significance level has been studied.[26] For this case, the authors concluded that if minimization adaptive stratification is used, an analysis of covariance should be employed. In order to obtain the proper significance level, the analysis should incorporate the same prognostic factors used in the randomization.

Response Adaptive Randomization

This procedure uses information on subject response to intervention during the course of the trial to determine the allocation of the next subject. Examples of response adaptive randomization models are the Play the Winner[27] and the Two-Armed Bandit[28] models. These models assume that the investigator is randomizing subjects to one of two interventions and that the primary response variable can be determined quickly relative to the total length of the study. Bailar[29] and Simon[30] review the uses of response adaptive stratification methods. Additional modifications or methods have been developed.[31,32]

The *Play the Winner* procedure may assign the first subject by the toss of a coin. The next subject is assigned to the same group as the first subject if the response to the intervention was a success; otherwise, the subject is assigned to the other group. That is, the process calls for staying with the winner until a failure occurs and then switching. The following illustrates a possible randomization scheme where S indicates intervention success and F indicates failure:

Example:

Assignment	Subject							
	1	2	3	4	5	6	7	8 ....
Group A	S	F				S	F	
Group B			S	S	F			S

Another response adaptive randomization procedure is the *Two-Armed Bandit* method which continually updates the probability of success as soon as the outcome for each subject is known. That information is used to adjust the probabilities of being assigned to either group in such a way that a higher proportion of future subjects would receive the currently "better" or more successful intervention.

Both of these response adaptive randomization methods have the intended purpose of maximizing the number of subjects on the "superior" intervention. They were developed in response to ethical concerns expressed by some clinical investigators about the randomization process. Although these methods do maximize the number of subjects on the "superior" intervention, the possible imbalance will almost certainly result in some loss of power and require more subjects to be enrolled into the study than would a fixed allocation with equal assignment probability.[33] A major limitation of these methods is that many clinical trials do not have an immediately occurring response variable. They also may have several response variables of interest with no single outcome easily identified as being the one upon which randomization should be based. Furthermore, these methods assume that the population from which the subjects are drawn is stable over time. If the nature of the study population should change and this is not accounted for in the analysis, the reported significance levels could be biased, perhaps severely.[34] Here, as before, the data analysis should ideally take into account the randomization process employed. For response adaptive methods, that analysis will be more complicated than it would be with simple randomization. Because of these disadvantages, response adaptive procedures are not commonly used.[35]

MECHANICS OF RANDOMIZATION

The manner in which the chosen randomization method is actually implemented is very important. If this aspect of randomization does not receive careful attention, the entire randomization process can easily be compromised, thus voiding any of the advantages for using it. To accomplish a valid randomization, it is recommended that an independent central unit be responsible for developing the randomization process and making the assignments of subjects to the appropriate group. For a single center trial, this central unit might be a statistician or clinician not involved with the care of the subjects. In the case of a multicenter trial, the randomization process is usually handled by the data or coordinating center. Ultimately, however, the integrity of the randomization process will rest with the investigator.

Chalmers and colleagues[36] reviewed the randomization process in 102 clinical trials, 57 where the randomization was unknown to the investigator and 45 where it was known. The authors reported that in 14% of the 57 studies, at least one baseline variable was not balanced between the two groups. For the studies with known randomization schedules, twice as many, or 26.7%, had at least one prognostic variable maldistributed. For 43 non-randomized studies, such imbalances occurred four times as often or in 58%. The authors emphasized that those recruiting and entering subjects into a trial should not be aware of the next intervention assignment.

In many cases when a fixed proportion randomization process is used, the randomization schedules are made before the study begins.[37-42] The investigators may call a central location, and the person at that location looks up the assignment for the next subject.[37] Another possibility, frequently used in trials involving acutely ill subjects, is to have a scheme making available sequenced and sealed envelopes containing the assignments.[38] As a subject enters the trial, he receives the next envelope in the sequence, which gives him his assignment. Envelope systems, however, are more subject to errors and tampering than the former method. In one study, personnel in a clinic opened the envelopes and arranged the assignments to fit their own preferences, accommodating friends and relatives entering the trial. In another case, an envelope fell to the bottom of the box containing the envelopes, thus changing the sequence in which they were opened. Many studies prefer the telephone system to protect against this problem. In an alternative procedure that has been used in several double-blind drug studies, medication bottles are numbered with a small perforated tab.[42] The bottles are distributed to subjects in sequence. The tab, which is coded to identify the contents, is torn off and sent to the central unit. This system is also subject to abuse unless an independent person is responsible for dispensing the bottles. Many clinical trials using a fixed proportion randomization schedule require that the investigator call the central location to verify that a subject is eligible to be in the trial before any assignment is made. This increases the likelihood that only eligible subjects will be randomized.

If adaptive randomization methods are used, the randomization process must be performed at the time the subject is ready to enter the trial. This is usually after a screening or baseline visit. White and Freedman[25] have implemented minimization methods using a special card index system. Small programmable computers could also be used.

Whatever system is chosen to communicate the intervention assignment to the investigator or the clinic, the intervention assignment should be given as closely as possible to the moment when both physician and subject are ready to begin the intervention. If the randomization takes place when the subject is first identified and the subject withdraws or dies before the intervention actually begins, a number of subjects will be randomized without participating in the study. An example of this occurred in a nonblinded trial of alprenolol in survivors of an acute myocardial infarction.[43] In that trial, 393 patients with a suspected myocardial infarction were randomized into the trial at the time of their admission to the coronary care unit. The alprenolol treatment was not initiated until two weeks later. Afterwards, 231 of the randomized subjects were excluded because a myocardial infarction could not be documented, death had occurred before therapy was begun, or various contraindications to therapy were noted. Of the 162 subjects who remained, 69 were in the alprenolol group and 93 were in the placebo group. This imbalance raises concerns over the comparability of the two groups and possible bias in reasons for subject exclusion. By delaying the randomization until initiation of therapy, the problem of these withdrawals could have been avoided.

RECOMMENDATIONS

For large studies involving more than several hundred subjects, the randomization should be blocked. If a large multicenter trial is being conducted, randomization should be stratified by center. Randomization stratified on the basis of other factors in large studies is usually not necessary, because randomization tends to make the study groups quite comparable for all risk factors. The subjects can still, of course, be stratified once the data have been collected and the study can be analyzed accordingly.

For small studies, the randomization should also be blocked, and stratified by center if more than one center is involved. Since the sample size is small, a few strata for important risk factors may be defined to assure that balance will be achieved for at least those factors. For a larger number of prognostic factors, the adaptive stratification techniques should be considered and the appropriate analyses performed. As in large studies, stratified analysis can be performed even if stratified randomization was not done. For many situations, this will be satisfactory.

APPENDIX

Adaptive Randomization Algorithm

Adaptive randomization can be used for more than two intervention groups, but for the sake of simplicity only two will be used here. In order to describe this procedure in more detail, a minimum amount of notation needs to be defined. First, let

x_{ik} = the number of subjects already assigned intervention k
$(k = 1,2)$ who have the same level of prognostic factor i
$(i = 1,2,...,f)$ as the new subject.

and define

$$x^t_{ik} = \begin{array}{ll} x_{ik} & \text{if } t \neq k \\ x_{ik} + 1 & \text{if } t = k \end{array}$$

The x^t_{ik} represents the change in balance of allocation if the new subject is assigned intervention t. Finally, let

$B(t)$ = function of the x^t_{ik}'s, which measures the "lack of balance" over all prognostic factors if the next subject is assigned intervention t.

Many possible definitions of $B(t)$ can be identified. As an illustrative example, let

$$B(t) = \Sigma^f_{i=1} \ w_i \ \text{Range} \ (x^t_{i1}, x^t_{i2})$$

where w_i = the relative importance of factor i to the other factors and the range is the absolute difference between the largest and smallest values of x^t_{i1} and x^t_{i2}.

The value of $B(t)$ is determined for each intervention ($t=1$ and $t=2$). The intervention with the smaller $B(t)$ is preferred, because allocation of the subject to that intervention will cause the least imbalance. The subject is assigned, with probability $P > \frac{1}{2}$, to the intervention with the smaller score, $B(1)$ or $B(2)$. The subject is assigned, with probability

$(1-P)$, to the intervention with the larger score. These probabilities introduce the random component into the allocation scheme. Note that if $P = 1$ and, therefore, $1-P = 0$, the allocation procedure is deterministic (no chance or random aspect) and has been referred to by the term "minimization."[24]

As a simple example of the adaptive stratification method, suppose there are two groups and two prognostic factors to control. The first factor has two levels and the second factor has three levels. Assume that 50 subjects have already been randomized and the following table summarizes the results.

Table A-1
Fifty Randomized Subjects by Group and Level of Factor (x_{ik}'s)*

Factor	1		2			
Level	1	2	1	2	3	
Group						Total
1	16	10	13	9	4	26
2	14	10	12	6	6	24
	30	20	25	15	10	50

*After Pocock and Simon.[12]

In addition, the function $B(t)$ as defined above will be used with the range of the x_{ik}'s as the measure of imbalance, where $w_1 = 3$ and $w_2 = 2$; that is, the first factor is 1.5 times as important as the second as a prognostic factor. Finally, suppose $P = ⅔$ and $1-P = ⅓$.

If the next subject to be randomized has the first level of the first factor and the third level of the second factor, then this corresponds to the first and fifth columns in the table. The task is to determine $B(1)$ and $B(2)$ for this subject as shown below.

a) Determine $B(1)$
   i) Factor 1, Level 1

	k	x_{1k}	x^1_{1k}	Range (x^1_{11}, x^1_{12})
	1	16	17	$17 - 14 = 3$
Group				
	2	14	14	

   ii) Factor 2, Level 3

	k	x_{2k}	x^1_{2k}	Range (x^1_{21}, x^1_{22})
	1	4	5	$5 - 6 = 1$
Group				
	2	6	6	

Using the formula given, $B(1)$ is computed as $3 \times 3 + 2 \times 1 = 11$.

b) Determine $B(2)$
   i) Factor 1, Level 1

	k	x_{1k}	x^2_{1k}	Range (x^2_{11}, x^2_{12})
	1	16	16	16 – 15 = 1
Group				
	2	14	15	

   ii) Factor 2, Level 3

	k	x_{2k}	x^2_{2k}	Range (x^2_{21}, x^2_{22})
	1	4	4	4 – 7 = 3
Group				
	2	6	7	

Then $B(2)$ is computed as $3 \times 1 + 2 \times 3 = 9$.

c) Now rank $B(1)$ and $B(2)$ from smaller to larger and assign with probability P the group with the smaller $B(t)$.

t	$B(t)$	Probability of Assigning t
2	$B(2) = 9$	$P = \frac{2}{3}$
1	$B(1) = 11$	$1 - P = \frac{1}{3}$

Thus, this subject is randomized to Group 2 with probability $\frac{2}{3}$ and to Group 1 with probability $\frac{1}{3}$. Note that if minimization were used ($P = 1$), the assignment would be Group 2.

REFERENCES

1. Armitage, P. The role of randomization in clinical trials. *Stat Med.* 1:345–352, 1982.
2. Byar, D.P., Simon, R.M., Friedewald, W.T., et al. Randomized clinical trials: perspectives on some recent ideas. *N Engl J Med.* 295:74–80, 1976.
3. Zelen, M. The randomization and stratification of patients to clinical trials. *J Chronic Dis.* 27:365–375, 1974.
4. Pocock, S.J. Allocation of patients to treatment in clinical trials. *Biometrics* 35:183–197, 1979.
5. Peto, R. Clinical trial methodology. *Biomedicine* 28 (special issue): 24–36, 1978.
6. Peto, R., Pike, M.C. Armitage, P. et al. Design and analysis of randomized clinical trials requiring prolonged observation of each patient. I. Introduction and design. *Br J Cancer* 34:585–612, 1976.

7. Brittain, E., and Schlesselman, J.J. Optimal allocation for the comparison of proportions. *Biometrics* 38:1003–1009, 1982.

8. Hill, A.B. The clinical trial. *Br Med Bull.* 71:278–282, 1951.

9. Matts, J.P., and McHugh, R.B. Analysis of accrual randomized clinical trials with balanced groups in strata. *J Chronic Dis.* 31:725–740, 1978.

10. Zelen, M. Aspects of the planning and analysis of clinical trials in cancer. Edited by J.N. Srivastava. In *A Survey of Statistical Design and Linear Models.* Amsterdam: North-Holland, 1975.

11. Coronary Drug Project Research Group. Factors influencing long term prognosis after recovery from myocardial infarction—three year findings of the Coronary Drug Project. *J Chronic Dis.* 27:267–285, 1974.

12. Pocock, S.J., and Simon, R. Sequential treatment assignment with balancing for prognostic factors in the controlled clinical trial. *Biometrics* 31:103–115, 1975.

13. Green, S.B., and Byar, D.P. The effect of stratified randomization on size and power of statistical tests in clinical trials. *J Chronic Dis.* 31:445–454, 1978.

14. Ducimetiere, P. Stratification. Edited by J.P. Boissel and C.R. Klimt. In *Multi-center Controlled Trials: Principals and Problems.* Paris: INSERM, 1979.

15. Simon, R. Restricted randomization designs in clinical trials. *Biometrics* 35:503–512, 1979.

16. Meier, P. Stratification in the design of a clinical trial. *Controlled Clin Trials* 1:355–361, 1981.

17. Grizzle, J.E. A note on stratifying versus complete random assignment in clinical trials. *Controlled Clin Trials* 3:365–368, 1982.

18. McHugh, R., and Matts, J. Post-stratification in the randomized clinical trial. *Biometrics* 39:217–225, 1983.

19. Fleiss, J.L. Multicentre clinical trials: Bradford Hill's contributions and some subsequent developments. *Stat Med.* 1:353–359, 1982.

20. Efron, B. Forcing a sequential experiment to be balanced. *Biometrika* 58:403–417, 1971.

21. Freedman, L.S., and White, S.J. On the use of Pocock and Simon's method for balancing treatment numbers over prognostic factors in the controlled clinical trial. *Biometrics* 32:691–694, 1976.

22. Begg, C.D., and Iglewicz, B. A treatment allocation procedure for sequential clinical trials. *Biometrics* 36:81–90, 1980.

23. Atkinson, A.C. Optimum biased coin designs for sequential clinical trials with prognostic factors. *Biometrika* 69:61–67, 1982.

24. Taves, D.R. Minimization: a new method of assigning patients to treatment and control groups. *Clin Pharmacol Ther.* 15:443–453, 1974.

25. White, S.J., and Freedman, L.S. Allocation of patients to treatment groups in a controlled clinical study. *Br J Cancer* 37:849–857, 1978.

26. Forsythe, A.B., and Stitt, F.W. Randomization or minimization in the treatment assignment of patient trials: validity and power of tests. Tech Rep No. 28, Health Science Computer Facility, University of California, Los Angeles, 1977.

27. Zelen, M. Play-the-winner rule and the controlled clinical trial. *J Am Stat Assoc.* 64:131–146, 1969.

28. Robbins, H. Some aspects of the sequential design of experiments. *Bull Am Math Soc.* 58:527–535, 1952.

29. Bailar, J.C. Patient assignment algorithms: an overview. In *Proceedings of the 9th International Biometric Conference.* Vol I, pp. 189–206. Raleigh, N.C.: The Biometric Society, 1976.

30. Simon, R. Adaptive treatment assignment methods and clinical trials. *Biometrics* 33:743–749, 1977.

31. Berry, D.A. Modified two-armed bandit strategies for certain clinical trials. *J Am Stat Assoc*. 73:339–345, 1978.

32. Bather, J.A. Randomized allocation of treatments in sequential experiments. *J R Stat Soc*. Series B 43:265–292, 1981.

33. Simon, R., Weiss, G.H. and Hoel, D.G. Sequential analysis of binomial clinical trials. *Biometrika* 62:195–200, 1975.

34. Simon, R., Hoel, D.G., and Weiss, G.H. The use of covariate information in the sequential analysis of dichotomous response experiments. *Commun Statist-Theor Meth*. 8:777–788, 1977.

35. Meier, P. Terminating a trial—the ethical problem. *Clin Pharmacol Ther*. 633–640, 1979.

36. Chalmers, T.C., Celano, P., Sacks, H.S., et al. Bias in treatment assignment in controlled clinical trials. *N Engl J Med*. 309:1358–1361, 1983.

37. Beta-Blocker Heart Attack Trial Research Group. A randomized trial of propranolol in patients with acute myocardial infarction. I. Mortality results. *JAMA* 247:1707–1714, 1982.

38. Hypertension Detection and Follow-up Program Cooperative Group. Five-year findings of the Hypertension Detection and Followup Program. Reduction in mortality of persons with high blood pressure, including mild hypertension. *JAMA* 242:2562–2571, 1979.

39. Aspirin Myocardial Infarction Study Research Group. A randomized controlled trial of aspirin in persons recovered from myocardial infarction. *JAMA* 243:661–669, 1980.

40. Multiple Risk Factor Intervention Trial Research Group. Multiple Risk Factor Interventional Trial. Risk factor changes and mortality results. *JAMA* 248:1465–1477, 1982.

41. CASS Principal Investigators and Their Associates. Coronary Artery Surgery Study (CASS): a randomized trial of coronary artery bypass surgery, survival data. *Circulation* 68:939–950, 1983.

42. Collaborative Group on Antenatal Steroid Therapy. Effect of antenatal dexamethasone administration on the prevention of respiratory distress syndrome. *Am J Obstet Gynecol*. 141:276–287, 1981.

43. Ahlmark, G., and Saetre, H. Long-term treatment with β-blockers after myocardial infarction. *Eur J Clin Pharmacol*. 10:77–83, 1976.

CHAPTER 6

Blindness

In any clinical trial bias is one of the main concerns. Bias may be defined as systematic error, or "difference between the true value and that actually obtained due to all causes other than sampling variability."[1] It can be caused by conscious factors, subconscious factors, or both. Bias can occur at a number of places in a clinical trial, from the initial design through data analysis and interpretation. The general solution to the problem of bias is to keep the subject and the investigator blinded, or masked, to the identity of the assigned intervention. One can blind several aspects of a trial: the identity of the intervention, and the assessment, classification and evaluation of the response variables.

FUNDAMENTAL POINT

To avoid potential problems of bias during data collection and assessment a clinical trial, ideally, should have a double-blind design. In studies where such a design is impossible, a single-blind approach and other measures to reduce potential bias are favored.

TYPES OF TRIALS
Unblinded

In an unblinded or open trial, both the subject and the investigator know to which intervention the subject has been assigned. Some kinds of trials can be conducted only in this manner. Such studies include those involving most surgical procedures, changes in lifestyle (eg, eating habits, exercise, cigarette smoking) or learning techniques.

An unblinded study is appealing for two reasons. First, all other things being equal, it is simpler to execute than other studies. The usual drug trial may be easier to design and carry out if blinding is not an issue. However, an unblinded trial need not be simple. For example, the Multiple Risk Factor Intervention Trial, which attempted to intervene on three primary risk factors for coronary heart disease simultaneously, was extraordinarily complex.[2] Second, investigators are likely to be more comfortable making decisions (such as whether or not to continue a subject on his assigned medication) if they know its identity.

The main disadvantage of an unblinded trial is the possibility of bias. Subject reporting of symptoms and side effects and prescription of concomitant or compensatory treatment are all susceptible to bias. (Other problems of biased data collection and assessment by the investigator are addressed in Chapter 10.) Subjects not on the new or experimental intervention may become dissatisfied and drop out of the trial in disproportionately large numbers. A trial of the possible benefits of ascorbic acid (vitamin C) in the common cold[3,4] started out as a double-blind study. However, it soon became apparent that many of the subjects, most of whom were hospital staff, discovered whether they were on ascorbic acid or placebo. Since evaluation of severity and duration of colds depended on the subject's reporting of his symptoms, this unblinding was important. Among those subjects who claimed not to know the identity of the treatment, ascorbic acid showed no benefit over placebo. In contrast, among subjects who knew or suspected what they were on, ascorbic acid did better than placebo. Therefore, preconceived notions about the benefit of a treatment, coupled with a subjective response variable, may have yielded biased reporting. Only the alertness of the investigators prevented them from arriving at probably false conclusions. In addition, as more subjects became aware of their medication's identity, the dropout rate increased. This was especially so in the placebo group.

Single-Blind

In a single-blind study, only the investigator is aware of which intervention each subject is receiving. The advantages of this design are similar to those of an unblinded study—it is usually simpler to carry out than a double-blind design, and knowledge of the intervention may help the investigator exercise his best judgment when caring for the subjects. Indeed, certain investigators are reluctant to participate in studies in which they do not know the study group assignment. They recognize that bias is partially reduced by keeping the subject blinded but feel that the subject's health and safety are best served if the investigator is not blinded.

The disadvantages of a single-blind design are similar to, though not so pronounced as, those of an unblinded design. The investigator avoids the problems of biased subject reporting, but he himself can affect the administration of non-study therapy, data collection and data assessment. For example, a single-blind study reported benefits from zinc administration in a group of people with taste disorders.[5] Because of the possibility of bias in a study using a response variable as subjective and hard to measure as taste, the study was repeated, using a type of cross-over, double-blind design.[6] This second study showed that zinc, when compared with placebo, did not relieve the taste disorders of the study group. The extent of the blinding of the subjects did not change; therefore, presumably, knowledge of drug identity by the investigator was important. The results of treatment cross-over were equally revealing. In the single-blind study, subjects who did not improve on placebo were placed on zinc. Improvement was then noted. However, in all four double-blind, cross-over procedures (placebo to zinc, placebo to placebo, zinc to zinc, zinc to placebo), the subjects who had previously shown no improvement did show benefit when given the second medication. Thus, the expectation that the subjects who failed to respond to the first drug were now being given an active drug may have been sufficient to produce a positive response.

Both unblinded and single-blind trials are vulnerable to another source of potential bias introduced by the investigators. The bias relates to group differences in compensatory and concomitant treatment. Investigators may feel that the control group is not being given the same opportunity as the intervention group and, as a result, may prescribe additional treatment as "compensation." This may be in the form of advice or therapy. For example, several studies have attempted blood pressure lowering as either the sole intervention, or as part of a broader effort. In each, the investigators made an intensive effort to persuade subjects in the intervention group to take their medication and to reduce their weight. To persuade successfully the investigators themselves had to be convinced that blood pressure reduction was beneficial. In other words, they had not suspended judgment regarding the possible efficacy of the intervention. When they were seeing subjects who had been assigned to the control group, this conviction was difficult to suppress. Therefore, subjects in the control group were likely to have been instructed about ways by which to lower their blood pressure. The result of compensatory treatment is a diminution of the difference between the intervention group and the "untreated," or control group. Working against this is the fact that investigators prefer to be associated with a study that gives positive findings. They may, therefore, subconsciously favor those in the intervention group when they deal with subjects, collect data, and assess results.

Concomitant treatment means any non-study therapy administered to subjects during a trial. If such treatment is likely to influence the response variable, this needs to be considered when determining sample size. Of more concern is the bias that can be introduced if concomitant treatment is applied unequally in the two groups. In order to bias the outcome of a trial, concomitant treatment must be effective, and it must be used in a high proportion of the subjects. When this is the case, bias is a possibility and may occur in either direction, depending on whether the concomitant treatment is preferentially used in the control, or in the intervention group. It is usually impossible to determine the direction and magnitude of such bias in advance or its impact after it has occurred.

In a trial of coronary artery bypass surgery versus medical treatment,[7] the number of subjects who smoked was equal in the two study groups at baseline. During follow-up, there were significantly fewer smokers in the surgical group than in the medical group. The effect of this group difference on the outcome of the trial is difficult, if not impossible to assess.

Double-Blind

In a double-blind study, neither the subjects nor the investigators responsible for following the subjects know the identity of the intervention assignment. Such designs are usually restricted to trials of drug efficacy. It is theoretically possible to design a study comparing two surgical procedures in which the surgeon performing the operation knows the type of surgery, but neither the subject nor the study investigator knows. Similarly, one might be able to design a study comparing two diets in which the food looks identical. However, such trials are uncommon.

The main advantage of a truly double-blind study is that the risk of bias is reduced. Preconceived ideas of the investigator will be less important, because he will not know which intervention a particular subject is receiving. Any effect of his actions, therefore, would theoretically occur equally in the intervention and control groups. As discussed later, the possibility of bias may never be completely eliminated. However, a well-designed and properly run double-blind study can minimize bias. As in the example of the trial of zinc and taste impairment, double-blind studies have at times led to results that differ from unblinded or single-blind studies. Such cases underline the importance of bias as a factor in clinical trials.

A case might be made that double-blind studies are more ethical than other studies. This is certainly so, if better and more useful information is obtained by means of this design. On the other hand, in a double-blind trial certain functions, which in open or single-blind studies could be accomplished by the investigators, must be taken over by others in order to maintain the blindness. Thus, an outside body needs to monitor the data for toxicity and benefit, especially in long-term trials. (Chapter 15 discusses data monitoring in greater detail.) A person other than the investigator who sees the subjects needs to be responsible for assigning the interventions to the subjects. The blinded data could also be analyzed by the person making assignments and presented to the monitoring body.

In many single- and double-blind drug trials the control group is placed on a placebo. Much debate in the past has centered on the ethics of using a placebo.[8] Conducting a placebo-control trial is justified if two situations pertain. First, there should be no standard intervention clearly superior to placebo. Second, the subject should fully understand that a placebo is being used and be aware of what his chances are of receiving either it or the alternative. Spodick[9] has succinctly pointed out that, despite persistent views to the contrary,[10] placebo therapy has at times been as effective as standard therapy. Even investigators who had initially considered use of a placebo as unethical have been persuaded of its value as a result of their clinical trial experience.[11]

Triple-Blind

A triple-blind study is an extension of the double-blind design; the committee monitoring response variables is not told the identity of the groups. The committee is simply given data for groups A and B. A triple-blind study has the theoretical advantage of allowing the monitoring committee to evaluate the response variable results more objectively. (This assumes that appraisal of efficacy and harm, as well as requests for special analyses, may be biased if group identity is known.) However, in a trial where the monitoring committee has an ethical responsibility to ensure subject safety, such a design may be counterproductive. When hampered in the safety-monitoring role, the committee cannot carry out its responsibility to minimize harm to the subjects. In addition, even if the committee could discharge its duties adequately while being kept blinded, many investigators would be uneasy participating in such a study. Though in most cases the monitoring committee looks only at group data and can rarely make informed judgments about individuals, the investigator still relies on the committee to safeguard his subjects. This may not be a completely rational approach because, by the time many monitoring committees receive data, often any emergency situation has long passed. Nevertheless, the discomfort many investigators feel about participating in double-blind studies would be magnified should the monitoring committee also be kept blinded.

Finally, people tend not to accept beneficial outcomes unless a statistically significant difference has been achieved. Rarely, though, will investigators continue a study in order to achieve a significant difference in an adverse direction; that is, until the intervention is statistically significantly worse than the control. Therefore, many monitoring committees demand to know which study groups are on which intervention.

SPECIAL PROBLEMS IN DOUBLE-BLIND
STUDIES

Double-blind studies are usually more complex and therefore more difficult to carry out than other trials. One must ensure that investigators remain blinded and that any data which conceivably might endanger blindness be kept from them during the study. An effective data-monitoring scheme must be set up, and emergency unblinding procedures must be established. These requirements pose their own problems and can increase the cost of a study. In the Aspirin-Myocardial Infarction Study,[12] a double-blind trial of aspirin in people with coronary heart disease, the investigators wished to monitor the action of aspirin on platelets. A postulated beneficial effect of aspirin relates to its ability to reduce the aggregation of platelets. Therefore, measuring platelet aggregation provided both an estimate of whether the aspirin-treated group was getting a sufficient dose and a basis for measurement of subject compliance. However, tests of platelet aggregation need to be performed shortly after the blood sample is drawn. The usual method is to have a laboratory technician insert the specimen in an aggregometer, add a material such as epinephrine (which, in the absence of aspirin, causes platelets to aggregate) and analyze a curve which is printed on a paper strip. In order to maintain the blind, the study needed to find a way to keep the technician from seeing the curve. Therefore, a cassette tape-recorder was substituted for the usual paper strip recorder and the indicator needle was covered. These changes required a modification of the aggregometer. All of the 30 clinics required this equipment, so the adjustment was expensive. However, it helped ensure the maintenance of the blind.

Drug studies, in particular, lend themselves to double-blind designs. One of the surest ways to unblind a drug study is to have dissimilar-appearing medications. When the treatment identity of one subject becomes known to the investigator, the whole trial is unblinded.

Naturally, subjects want to be on the "better" intervention. In a drug trial, the "better" intervention usually is presumed to be the new one; in the case of a placebo-control trial it is presumed to be the active medication. Investigators may also be curious about a drug's identity. For these reasons, consciously or unconsciously, both subjects and investigators may try to unblind the medication. Unblinding can be done deliberately by going so far as to have the drug analyzed, or in a less purposeful manner by "accidentally" breaking open capsules, holding pills up to the light, carefully testing them, or by making any of numerous other tests. In the first case, which may have occurred in the vitamin C study discussed earlier, little can be done to ensure blinding absolutely. Curious subjects and investigators can discover numerous ways to unblind the trial, whatever precautions are taken. Probably, however, the less purposeful unblinding is more common.

Matching of Drugs

Proper matching has received little attention in the literature. Most reports of drug studies do not indicate how closely tablets or capsules resembled one another, or how great a problem was caused by imperfect matching. However, one report[13] is disturbing. The authors noted that, of 22 studies surveyed, only five had excellent matching between the drugs being tested. A number of features of matching must be considered.[14] Cross-over studies, where each subject sees both medications, require the most care in matching. Visual discrepancies can occur in size, shape, color and texture. Ensuring that these characteristics are identical may not be simple. In the case of tablets, dyes or coatings may adhere differently to the active ingredient than to the placebo, causing slight differences in color or sheen. Agents can also differ in odor. The taste and the local action on the tongue of the active medication are likely to be different than those of the placebo. For example, propranolol is a topical anesthetic which causes lingular numbness if held in the mouth.

Differences may become evident only after some time, due to degradation of the active ingredient. Freshly prepared aspirin is relatively odor free, but after a while, tell-tale acetic acid accumulates. Less obviously, the weight or specific gravity of the tablets may differ. Matching the agents on all of these characteristics may be impossible. However, if a great deal of effort and money are being spent on the trial, a real attempt to ensure matching makes sense. The investigator also needs to make sure that the containers are identical. Bottles and vials need to be free of any marks other than codes which are indecipherable except with the key.

Drug preparation should be pretested if it is possible. One method is to have a panel of observers unconnected with the study compare samples of the medications. Perfect matches are almost impossible to obtain and some differences are to be expected. However, beyond detecting differences, it is important to assess whether the observers can actually identify the agents. If not, slightly imperfect matches may be tolerated. The investigator must remember that, except in cross-over studies, the subject has only one drug and is therefore not able to make a comparison. On the other hand, subjects may meet and talk in waiting rooms, or in some other way compare notes or pills. Of course, staff always have the opportunity to compare different preparations and undermine the integrity of a study.

Use of substances to mask characteristic taste, color, or odor is often advocated. Adding vanilla to the outside of tablets may mask an odor; adding dyes will mask dissimilar colors. A substance such as quassin will impart a bitter taste to the preparations. Not only will quassin mask differences in taste, but it will also effectively discourage subjects from biting into a preparation more than once. However, the possibility that chemical substances may have toxic effects after long-term use must be considered. It is usually prudent to avoid using extra substances unless absolutely essential to prevent unblinding of the study.

Rarely do publications of trial results discuss possible inadequate matching. An exception is the vitamin C study[3,4] which suffered from a breakdown of the double-blind. One possible reason given by the investigators was that, in the rush to begin the study, the contents of the capsules were not carefully produced. The lactose placebo could easily be distinguished from ascorbic acid by taste, as the study subjects quickly discovered.

Coding of Drugs

By drug coding is meant the labeling of individual drug bottles or vials so that the identity of the drug is not disclosed. Coding is usually done by means of assigning a random set of numbers to the active drug and a different set to the control. As many different drug codes as are logistically feasible should be used. At least in smaller studies, each subject should have a unique drug code which remains with him for the duration of the trial. If only one code were used for each study group, unblinding a single subject would result in unblinding everybody. Furthermore, many drugs have specific side effects. One side effect in one subject may not be attributable to the drug, but a constellation of several side effects in several subjects with the same drug code may easily unblind the study.

Unfortunately, in large studies it becomes cumbersome to make up and stock drugs under many different codes. This is true for the investigator dispensing the medication as well as for the person labeling and coding. When several subjects have the same code, there is less chance that a particular subject will run out of medication, because in emergencies, drugs can be borrowed from another subject's supply. An investigator needs to balance these logistic concerns against the risks of unblinding when he determines the number of drug codes.

Unblinding

The phrase "truly double-blind study" was used earlier in the text. While many studies are designed as double- or single-blind, it is unclear how many, in fact, are truly blind. Certainly, drugs have side effects, some of which are fairly characteristic. Existence or absence of side effects does not necessarily unblind drug assignment, since all people on drugs do not develop reactions and some people on placebo develop events which can be mistaken for drug side effects. It is well known that aspirin is associated with gastrointestinal problems. In the Coronary Drug Project Aspirin Study,[15] 5.4% of the subjects in the aspirin group developed gastritis. On the other hand, 3.9% of the placebo subjects had the same complaints. To have made definite claims regarding study group identity based on presence or absence or gastritis would have led to many errors.

Occasionally, accidental unblinding occurs. In some studies, a special center labels and distributes drugs to the clinic where subjects are seen. Obviously, each carton of drugs sent from the pharmaceutical company to this distribution center must contain a packing slip identifying the drug. The distribution center puts coded labels on each bottle and removes the packing slip before sending the drugs to the investigator. In one instance, one carton contained two packing slips by mistake. The distribution center, not realizing this, shipped the carton to the investigator with the second packing slip enclosed. Thus, it is advisable to empty cartons completely before re-using them.

Laboratory errors have also occurred. These are particularly likely when, to prevent unblinding, only some laboratory results can be given to the investigators. Occasionally investigators have received the complete set of laboratory results. This usually happens at the beginning of a study before "bugs" have been worked out, or when the laboratory hires new personnel who are unfamiliar with the procedures. If a commercial laboratory performs the study determinations, the tests should be done in a special area of the laboratory, with safeguards to prevent study results from getting intermingled with routine work.

In addition, monitoring the use of study medication prescribed outside the study is essential. Any group differences might be evidence of a deficiency in the blind. Another way of estimating the success of a double-blind design is to monitor specific intermediate effects of the study medication. The use of platelet aggregation in the Aspirin-Myocardial Infarction Study is an example. An unusually large number of subjects with non-aggregating platelets in the placebo group would raise the suspicion that the blind had been broken.

Official breaking of the blind may be necessary. There are bound to be situations that require disclosures, especially in long-term studies. Perhaps the study drug requires tapering the dosage. In an emergency, knowledge that a subject is or is not on the active drug would indicate whether tapering is necessary. Usually, most emergencies can be handled by withdrawing the medication without breaking the blind. When the treating physician is different from the study investigator, a third party can obtain the blinded information from the pharmacy or central data repository and relate the information to the treating physician. In this way, the subject and the investigator need not be unblinded. When unblinding does occur, the investigator should review the circumstances which led to it. Knowledge of the kind of intervention seldom influences emergency care of the subject, and such reviews have helped reduce the frequency of further unblinding.

A procedure should be developed to break the blind quickly for any individual subject at any time should it be in his best interest. Such systems include having labels on file in the hospital pharmacy or other accessible locations, or having "on call" 24 hours a day someone who can decode the assignment. In order to avoid needless breaking of the code, someone other than the investigator could hold a list that reveals the identity of each drug code. Alternatively, each study medication bottle label might have a sealed tear-off portion that would be filed in the pharmacy or with the subject's records. In an emergency, the seal could be torn and the drug identity revealed. Care should be taken to ensure that the sealed portion is of appropriate color and thickness to prevent reading through it. In one study, the sealed labels attached to the medication bottles were transparent when held up to strong light.

In summary, double-blind trials require careful planning and constant monitoring to ensure that the blind is maintained and that subject safety is not jeopardized.

Assessment of Blindness

After a study is completed, estimating the degree to which the blind was maintained is worthwhile. One way of estimating before officially disclosing drug identity is to ask the subject and the clinic staff to guess to which group the subject was assigned. Ideally, in a trial with one-half of the subjects on active medication and one-half on control, the guesses should be correct 50% of the time in each group. To the degree that 50% is exceeded, the amount of unblinding can be estimated. If substantially fewer than half of the guesses in each group are correct, one might suspect that people did know but were trying not to admit it.

Shortly before the end of the Aspirin Myocardial Infarction Study, 380 of the 4524 subjects were asked whether they knew if they were taking aspirin or placebo.[16] Slightly over half correctly identified what they were taking, a little over one-fourth chose the wrong agent, and the remainder either refused to guess or selected a drug not being tested. Most of those who guessed correctly were not certain of their selection. Those who made special efforts to break the blind did better. Compliance to the study medication did not seem to be related to guessing correctly. Even though blinding was found to be imperfect, it clearly contributed to the integrity of the trial.

REFERENCES

1. Mausner, J.S. and Bahn, A.K. *Epidemiology: An Introductory Text*. Philadelphia: W.B. Saunders, 1974.

2. Multiple Risk Factor Intervention Trial Research Group. Multiple Risk Factor Intervention Trial: risk factor changes and mortality results. *JAMA* 248: 1465–1477, 1982.

3. Karlowski, T.R., Chalmers, T.C., Frenkel, L.D. et al. Ascorbic acid for the common cold: a prophylactic and therapeutic trial. *JAMA* 231:1038–1042, 1975.

4. Lewis, T.L. Karlowski, T.R. Kapikian, A.Z. et al. A controlled clinical trial of ascorbic acid for the common cold. *Ann NY Acad Sci*. 258:505–512, 1975.

5. Schechter, P.J., Friedewald, W.T., Bronzert, D.A. et al. Idiopathic hypogensia: a description of the syndrome and a single-blind study with zinc sulfate. *Int Rev Neurobiol*. (suppl I): 125–140, 1972.

6. Henkin, R.I., Schechter, P.J., Friedewald, W.T. et al. A double-blind study of the effects of zinc sulfate on taste and smell dysfunction. *Am J Med Sci*. 272:285–299, 1976.

7. European Coronary Surgery Study Group. Coronary-artery bypass surgery in stable angina pectoris: survival at two years. *Lancet* 1:889–893, 1979.

8. Bok, S. The ethics of giving placebos. *Sci Am*. 231:17–23, (Nov) 1974.

9. Spodick, D.H. Letter to the editor. *N Engl J Med*. 292:653, 1975.

10. Wilks, H.S. Letter to the editor. *N Engl J Med*. 292:321, 1975.

11. Prout, G.R., Jr., Bross, I.D.J., Slack, N.H., and Ausman, R.K. Carcinoma of the bladder. 5-fluorouracil and the critical role of a placebo. *Cancer* 22:926–931, 1968.

12. Aspirin Myocardial Infarction Study Research Group. A randomized controlled trial of aspirin in persons recovered from myocardial infarction. *JAMA* 243:661–669, 1980.

13. Hill, L.E., Nunn, A.J., and Fox, W. Matching quality of agents employed in "double-blind" controlled clinical trials. *Lancet* 1:352–356, 1976.

14. Joyce, C.R.B. Psychological factors in the controlled evaluation of therapy. Edited by E. Balint. In *Psychopharmacology: Dimensions and Perspectives*. London: Tavistock Publications, 1968.

15. Coronary Drug Project Research Group. Aspirin in coronary heart disease. *J Chronic Dis*. 29:625–642, 1976.

16. Howard, J., Whittemore, A.S., Hoover, J.J., et al. How blind was the patient blind in AMIS? *Clin Pharmacol Ther*. 32:543–553, 1982.

CHAPTER 7

Sample Size

The size of the study should be considered early in the planning phase. Often no formal sample size is ever calculated. Instead, the number of subjects available to the investigators during some period of time determines the size of the study. Many clinical trials which do not carefully consider the sample size requirements turn out to lack the power or ability to detect intervention effects of fairly substantial magnitude and clinical importance. Freiman and colleagues[1] reviewed the power of 71 published randomized controlled clinical trials which failed to find significant differences between groups. "Sixty-seven of the trials had a greater than 10% risk of missing a true 25% therapeutic improvement, and with the same risk, 50 of the trials could have missed a 50% improvement." The danger in studies with low statistical power is that interventions which could be beneficial are discarded without adequate testing and may never be considered again.

This chapter presents a general overview of sample size estimation with some details. Lachin has written a more technical discussion of this topic.[2] The medical literature also contains other references concerning sample size estimation.[3-5]

FUNDAMENTAL POINT

Clinical trials should have sufficient statistical power to detect differences between groups considered to be of clinical interest. Therefore, calculation of sample size with provision for adequate levels of significance and power is an essential part of planning.

Before a discussion of sample size and power calculations, it must be emphasized that, for several reasons, a sample size calculation provides only an estimate of the needed size of a study.[3] First, parameters used in the calculation are estimates, and as such, have an element of uncertainty. Second, the estimate of the relative effectiveness of the intervention over the control may be based on a population different from that intended to be studied. Third, the estimated effectiveness is often overly optimistic. Fourth, during the final planning stage of a trial, revisions of inclusion and exclusion criteria may influence the type of subject entering the study and thus alter earlier assumptions used in the sample size calculation. Assessing the impact of such changes in criteria is usually impossible. Experience indicates that subjects enrolled into control groups do better than the population from which the subjects were drawn. The reasons are not entirely clear. One factor could be that subjects with the highest risk of developing the outcome of interest are excluded in the screening process. In trials involving chronic diseases, because of the research protocol, subjects might receive more care and attention than they would normally be given, thus improving their prognosis. Subjects assigned to the control group may, therefore, be better off than if they had not been in the trial at all. Finally, sample size calculations are based on mathematical models which may only approximate the true, but unknown, distribution of the response variables.

Due to the approximate nature of sample size calculations, the investigator should be as conservative as can be justified while still being realistic in estimating the parameters used in the calculation. If a sample size is drastically overestimated, the trial may be judged as unfeasible. If the sample size is underestimated, there is a good chance the trial will fall short of demonstrating any effectiveness of the intervention. In general, as long as the calculated sample size is realistically obtainable, it is better to overestimate the size and possibly terminate the trial early (Chapter 15) than to underestimate and need to justify an increase or an extension in follow-up, or worse, to arrive at incorrect conclusions.

STATISTICAL CONCEPTS

An understanding of the basic statistical concepts of hypothesis testing, significance level, and power is essential for a discussion of sample size estimation. A brief review of these concepts follows. Further discussion can be found in basic medical statistics textbooks[6-11] as well as selected papers.[1,3,12]

Except where indicated, studies of one intervention group and one control group will be discussed. With some adjustments, sample size calculations can be made for studies with more than two groups. Before computing sample size, the primary response variable used to judge the effectiveness of intervention must be identified. This chapter will consider three basic kinds: 1) dichotomous response variables, such as success and failure, 2) continuous response variables, such as blood pressure level or a change in blood pressure, and 3) time to failure (or occurrence of a clinical event).

For the dichotomous response variables, the event rates in the intervention group (p_I) and the control group (p_C) are compared. For continuous response variables, the true but unknown mean level in the intervention group (μ_I) is compared with the mean level in the control group (μ_C). For survival data, a hazard rate, λ, is often compared for the two study groups or at least is used for sample size estimation. Sample size estimates for response variables which do not exactly fall into any of the three categories can usually be approximated by one of them.

In terms of the primary response variable, p_I will be compared with p_C or μ_I will be compared with μ_C. This discussion will use only the event rates, p_I and p_C, although the same concepts will hold if response levels μ_I and μ_C are substituted appropriately. Of course, the investigator does not know the true values of the event rates. The clinical trial will give him only estimates of the event rates, denoted by $\hat{p}_I$ and $\hat{p}_C$. Typically, an investigator tests whether or not a true difference exists between the event rates of subjects in the two groups. The traditional way of stating this is in terms of a null hypothesis, denoted H_0, which states that no difference between the true event rates exists $(H_0: p_C - p_I = 0)$. The goal is to test H_0 and decide whether or not to reject it. That is, the null hypothesis is assumed to be true until proven otherwise.

Since only estimates of the true event rates are obtained, it is possible that, even if the null hypothesis is true $(p_C - p_I = 0)$, the observed event rates might by chance be different. If the observed differences in event rates are large enough by chance alone, the investigator might reject the null hypothesis incorrectly. This false positive occurrence or Type I error should be made as few times as possible. The probability of this Type I error is called the significance level and is denoted by α. The probability of observing differences as large as, or larger than the difference actually observed given that H_0 is true is called the "P-value," denoted as P. The decision will be to reject H_0 if $P \leqslant \alpha$. While the level of α chosen is somewhat arbitrary, the ones used and accepted traditionally are 0.01 or 0.05. As will be shown later, as α is set smaller, the required sample size increases.

If the null hypothesis is not in fact true, then another hypothesis, called the alternative hypothesis, denoted by H_A, must be true. That is, the true difference between the event rates p_I and p_C is some value δ where $\delta \neq 0$. The observed difference, $\hat{p}_C - \hat{p}_I$, can be quite small by chance alone even if the alternative hypothesis is true. Therefore, the investigator could, on the basis of small observed differences, fail to reject H_0 when he should. This is called a Type II error or a false negative result. The probability of a Type II error is denoted by β. The value of β is dependent on the specific value of δ, the true but unknown difference in event rates between the two groups, as well as on the sample size and α. The probability of correctly rejecting H_0 is denoted $1-\beta$ and is called the power of the study. Power quantifies the ability of the study to find true differences of various values δ. Since β is a function of α, the sample size and δ, 1-β is also a function of these parameters. The plot of 1-β versus δ for a given sample size is called the power curve and is depicted in Figure 7-1. On the horizontal axis, values of δ are plotted from 0 to an upper value, δ_A (0.25 in this figure). On the vertical axis, the probability or power of detecting a true difference δ is shown for a given significance level and sample size. In constructing this specific power curve, a sample size of 100 in each group, a one-sided significance level of 0.05 and a control group event rate of 0.5 (50%) were assumed. Note that as δ increases, the power to detect δ also increases. For example, if $\delta = 0.10$ the power is approximately 0.40. When $\delta = 0.20$ the power increases to about 0.90. Typically, investigators like to have a power (1-β) of around 0.90 or 0.95 when planning a study; that is to have a 90% or 95% chance of finding a statistically significant difference between the event rates, given that a difference, δ, actually exists.

Since the significance level α should be small, say 0.05 or 0.01, and the power (1-β) should be large, say 0.90 or 0.95, the only quantities which are left to vary are δ and the total sample size. In planning a clinical trial, the investigator hopes to detect a difference of specified magnitude δ or larger. One factor that enters into the selection of δ is the minimum difference between groups that is judged to be clinically important. In addition, previous research may provide estimates of δ. The exact nature of the calculation of the sample size, given α, 1-β and δ is considered here. It can be assumed that the randomization strategy will allocate an equal number (N) of subjects to each group. If the variability in the responses for the two groups is approximately the same, equal allocation provides a more powerful design than does unequal allocation. For unequal allocation to yield an appreciable increase in power, the variability needs to be substantially different in the groups.[13] Since equal allocation is usually easier to implement, it is the more frequently used strategy.

Before a sample size can be calculated, classical statistical theory says that the investigator must decide whether he is interested in differences in one direction only (one-sided test) — say improvements in intervention over control — or in differences in either direction (two-sided test). This latter case would represent testing the hypothesis that the new intervention is either better or worse than the control. In general, two-sided tests should be used unless there is strong justification for expecting a difference in only one direction. An investigator should always keep in mind that any new intervention could be harmful as well as helpful. However, as discussed in Chapter 15, some investigators may not be willing to prove the intervention harmful and would terminate a study if the results are only suggestive of harm.

If a one-sided test of hypothesis is chosen, in most circumstances the significance level ought to be half what the investigator would use for a two-sided test. For example, if .05 is the two-sided significance level typically used, .025 would be used for the one-sided test. This requires the same degree of evidence or scientific documentation to declare a treatment effective, regardless of the one-sided vs. two-sided question.

As mentioned above, the total sample size $2N$ is a function of the significance level (α), the power ($1-\beta$) and the size of the difference in response (δ) which is to be detected. Changing either α, $1-\beta$ or δ will result in a change in $2N$. As the magnitude of the difference δ decreases, the larger the sample size must be to guarantee a high probability of finding that difference. If the calculated sample size is larger than can be realistically obtained, then one or more of the parameters in the design must be modified. Since the significance level is usually fixed at 0.05 or 0.01, the investigator should generally reconsider the value selected for δ and increase it, or keep δ the same and settle for a less powerful study. If neither of these alternatives is satisfactory, serious consideration should be given to abandoning the trial.

Rothman[14] has argued that journals should encourage using confidence intervals to report clinical trial results instead of significance levels. McHugh and Le[15] discuss sample size formulae from this approach. The methods presented here do not preclude the presentation of results as confidence intervals and, in fact, investigators ought to do so.

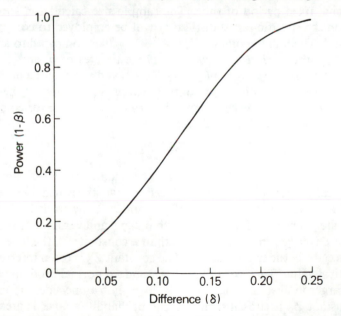

Figure 7-1 A power curve for increasing differences (δ) between the control group rate of 0.5 and the intervention group rate with a one-sided significance level of 0.05 and a total sample size ($2N$) of 200.

However, unless there is an awareness of the relationship between two approaches, as McHugh and Le have pointed out, the confidence interval method might yield a power of only 50% to detect a specified difference. Thus, some care needs to be taken in using this method.

So far, it has been assumed that the data will be analyzed only once at the end of the trial. However, as discussed in Chapter 15, the response variable data may be reviewed periodically during the course of a study. Thus, the probability of finding significant differences by chance alone is increased.[16] This means that the significance level α may need to be adjusted to compensate for the increase in the probability of a Type I error. For purposes of this discussion, assume that α carries the usual values of 0.05 or 0.01.

The sample size calculation should also employ the statistic which will be used in data analysis. Thus, there are many sample size formulations. Methods that have proven useful will be discussed.

SAMPLE SIZE CALCULATIONS
FOR DICHOTOMOUS RESPONSE VARIABLES

Suppose the primary response variable is the occurrence of an event over some fixed period of time. The sample size calculation should be based on the specific test statistic that will be employed to compare the outcomes. The null hypothesis $H_0 (p_C - p_I = 0)$ is compared to an alternative hypothesis $H_A (p_C - p_I \neq 0)$. The estimates of p_I and p_C are $\hat{p}_I$ and $\hat{p}_C$ where $\hat{p}_I = r_I/N_I$ and $\hat{p}_C = r_C/N_C$, r_I and r_C being the number of events in the intervention and control groups and N_I and N_C being the number of subjects in each group. The usual test statistic for comparing such dichotomous or binomial responses is

$$Z = (\hat{p}_C - \hat{p}_I)/\sqrt{\bar{p}(1-p)(1/N_C + 1/N_I)}$$

where $\bar{p} = (r_I + r_C)/(N_I + N_C)$. The square of the Z statistic is algebraically equivalent to the chi-square statistic which is often employed as well. For large values of N_I and N_C, the statistic Z has approximately a normal distribution with mean 0 and variance 1. If the test statistic Z is larger in absolute value than a constant Z_α, the investigator will reject H_0 in the two-sided test. The constant Z_α is often referred to as the critical value. The probability of a standard normal random variable being larger in absolute value than Z_α is α. For a one-sided hypothesis, the constant Z_α is chosen such that the probability that Z is greater (or less) than Z_α is α. For a given α, Z_α is larger for a two-sided test than for a one-sided test (Table 7-1). Z_α for a two-sided test with $\alpha = 0.10$ has the same value as Z_α for a one-sided test with $\alpha = 0.05$.

The sample size required for the design to have a significance level α and a power of 1-β to detect true differences of at least δ between the event rates p_I and p_C can be expressed by the formula[9]:

$$2N = 2\{ Z_\alpha \sqrt{2\,\overline{p}(1-\overline{p})} + Z_\beta \sqrt{p_C(1-p_C) + p_I(1-p_I)}\}^2/(p_C - p_I)^2$$

where $2N$ = total sample size (subjects) with $\overline{p} = (p_C + p_I)/2$; Z_α is the critical value which corresponds to the significance level α; and Z_β is the value of the standard normal value not exceeded with probability β. Z_β corresponds to the power 1-β (eg, if 1-β = 0.90, Z_β = 1.282). Values of Z_α and Z_β are given in Tables 7-1 and 7-2 for several values of α and 1-β. Note that the definition of $\overline{p}$ given earlier is equivalent to the definition of $\overline{p}$ given here when $N_I = N_C$; that is, when the two study groups are of equal size.

As an example, suppose the annual event rate in the control group is anticipated to be 20%. The investigator hopes that the intervention will reduce the annual rate to 15%. The study is planned so that each subject will be followed for two years. Therefore, if the assumptions are accurate, approximately 40% of the subjects in the control group and 30% of

Table 7-1
Z_α for Sample Size Formulas for Various Values of α

| | Z_α | |
α	One-sided Test	Two-sided Test
0.10	1.282	1.645
0.05	1.645	1.960
0.025	1.960	2.240
0.01	2.326	2.576

Table 7-2
Z_β for Sample Size Formulas for Various Values of Power (1-β)

1-β	Z_β
0.50	0.00
0.60	0.25
0.70	0.53
0.80	0.84
0.85	1.03
0.90	1.282
0.95	1.645
0.975	1.960
0.99	2.326

the subjects in the intervention group will develop an event. Thus, the investigator sets $p_C = 0.40$, $p_I = 0.30$, and, therefore, $\bar{p} = (0.4 + 0.3)/2 = 0.35$. The study is designed as two-sided with a 5% significance level and 90% power. From Tables 7-1 and 7-2, the two-sided 0.05 critical value is 1.96 for Z_α and 1.282 for Z_β. Substituting these values into the right-hand side of the sample size formula yields

$$2\{1.96\sqrt{2(0.35)(0.65)} + 1.282\sqrt{0.4(0.6) + 0.3(0.7)}\}^2/(0.4 - 0.3)^2$$

Evaluating this expression, $2N = 952.3$. Therefore, the calculated total sample size is 960, rounding up to the nearest 10, or 480 in each group.

Sample size estimates using this formula are given in Table 7-3 for a variety of values of p_I and p_C, for both one-sided and two-sided tests, and for $\alpha = 0.05$ and $1-\beta = 0.80$ or 0.90. For the example just considered with $\alpha = 0.05$ (two-sided), $1-\beta = 0.90$, $p_C = 0.4$ and $p_I = 0.3$, the total sample size using Table 7-3 is 960. This table shows that, as the difference in rates between groups increases, the sample size decreases.

Table 7-3
Approximate* Total Sample Size for Comparing Various Proportions in Two Groups with Significance Level (α) of 0.05 and Power ($1-\beta$) of 0.80 and 0.90

True Proportions		$\alpha = 0.05$ (one-sided)		$\alpha = 0.05$ (two-sided)	
p_C (Control Group)	p_I (Intervention Group)	$1-\beta$ 0.90	$1-\beta$ 0.80	$1-\beta$ 0.90	$1-\beta$ 0.80
0.60	0.50	850	610	1040	780
	0.40	210	160	260	200
	0.30	90	70	120	90
	0.20	50	40	60	50
0.50	0.40	850	610	1040	780
	0.30	210	150	250	190
	0.25	130	90	160	120
	0.20	90	60	110	80
0.40	0.30	780	560	960	720
	0.25	330	240	410	310
	0.20	180	130	220	170
0.30	0.20	640	470	790	590
	0.15	270	190	330	250
	0.10	140	100	170	130
0.20	0.15	1980	1430	2430	1810
	0.10	440	320	540	400
	0.05	170	120	200	150
0.10	0.05	950	690	1170	870

*Sample sizes are rounded up to the nearest 10.

As the power (1-β) increases, the sample size also increases. Finally, although not shown in the Table, as the significance level α decreases, the sample size increases.

The event rate in the intervention group can be written as $p_I = (1-k)p_C$ where k represents the proportion that the control group event rate is expected to be reduced by the intervention. Figure 7-2 shows the total sample size $2N$ versus k for several values of p_C using a two-sided test with $\alpha = 0.05$ and 1-$\beta = 0.90$. In the example where $p_C = 0.4$ and $p_I = 0.3$, the intervention is expected to reduce the control rate by 25% or $k = 0.25$. In Figure 7-2 locate $k = 0.25$ on the horizontal axis and move up vertically until the curve labeled $p_C = 0.4$ is located. The point on this curve corresponds to a $2N$ of approximately 960. Notice that as the control group event rate p_C decreases, the sample size required to detect

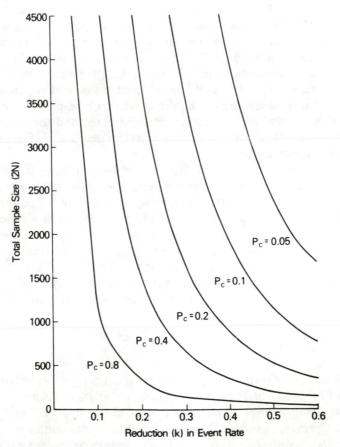

Figure 7-2 Relationship between total sample size (2N) and reduction (k) in event rate for several control group event rates (pc), with a two-sided significance level of 0.05 and power of 0.90.

the same proportional reduction increases. Studies with small event rates (eg, $p_C = 0.1$) require large sample sizes unless the interventions are dramatically effective.

In order to make use of the sample size formula or table, it is necessary to know something about P_C and k. The estimate for P_C is usually obtained from a previous study of similar subjects. In addition, the investigator must choose k based on preliminary evidence of the potential effectiveness of the intervention or be willing to specify some minimum difference or reduction that he wants to detect. Obtaining this information is difficult in many cases. Frequently, estimates may be based on a small amount of data. In such cases, several sample size calculations based on a range of estimates help to assess how sensitive the sample size is to the uncertain estimates of p_C, k or both. The investigator may want to be conservative and take the largest, or nearly largest, estimate of sample size to be sure his study has sufficient power. The power $(1-\beta)$ for various values of δ can be compared for a given sample size $2N$, significance level α, and control rate p_C. By examining a power curve such as in Figure 7-1, it can be seen what power the trial has for detecting various differences in rates, δ. If the power is high—say 0.80 or larger—for the range of values δ that are of interest, the sample size is probably adequate. The power curve can be especially helpful if the number of available subjects is relatively fixed and the investigator wants to assess the probability that the trial can detect any of a variety of reductions in event rates.

Investigators often overestimate the number of eligible subjects who can be enrolled in a trial. The actual number enrolled may fall short of goal. To examine the effects of smaller sample sizes on the power of the trial, the investigator may find it useful to graph power as a function of various sample sizes. If the power falls far below 0.8 for a sample size that is very likely to be obtained, he can expand the recruitment effort, hope for a larger intervention effect than was originally assumed, accept the reduced power and the consequences or abandon the trial.

To determine the power, the sample size equation is solved for Z_β.

$$Z_\beta = \frac{-Z_\alpha \sqrt{2p(1-p)} + \sqrt{N}(p_C - p_I)}{\sqrt{p_C(1-p_C) + p_I(1-p_I)}}$$

where $\bar{p}$ as before is $(p_C + p_I)/2$. The term Z_β can be translated into a power of $1-\beta$ by use of Table 7-2. For example, let $p_C = 0.4$ and $p_I = 0.3$. For a significance level of 0.05 in a two-sided test of hypothesis, $Z_\alpha = 1.96$. In a previous example, it was shown that a total sample of approximately 960 subjects or 480 per group is necessary to achieve a power of 0.90. Substituting $Z_\alpha = 1.96$, $N = 480$, $p_C = 0.4$ and $p_I = 0.3$, 1.295 is obtained. The closest value of Z_β in Table 7-2 is 1.282 which corresponds to a power of 0.90. (If the exact value of $N = 476$ were used, the value of Z_β would be 1.282.) Suppose an investigator thought he could get only 350 subjects per group instead of the estimated 480. Then $Z_\beta = 0.818$ which means that the power is somewhat less than 0.80. If the value of Z_β is negative, the power is less than 0.50. For more details of power calculations, a standard text in biostatistics[7-11] should be consulted.

For a given $2N$, α, $1\text{-}\beta$, and p_C the reduction in event rate that can be detected can also be calculated. This function is nonlinear and, therefore, the details will not be presented here. Approximate results can be obtained by scanning Table 7-3, by using the formula for several p_I's until the sample size approaches the planned number, or by using a figure where sample sizes have been plotted. In Figure 7-2, $\alpha = 0.05$ and $1\text{-}\beta = 0.90$. If sample size is selected as 1000, with $p_C = 0.4$, k is determined to be about 0.25. This means that the expected p_I would be 0.3. As can be seen in Table 7-3, the actual sample size for these assumptions is 960.

The above approach yields an estimate which is more accurate as the sample size increases. Modifications[17-24] have been developed which give some improvement in accuracy to the approximate formula presented for small studies. However, given that sample size estimation is somewhat imprecise due to assumptions of intervention effects and event rates, the formulation presented is probably adequate for most studies.

Adjusting Sample Size to Compensate for Noncompliance to Intervention

During the course of a clinical trial, subjects will not always comply with their prescribed intervention schedule. The reason is often that the subject cannot tolerate the dosage of the drug or the degree of intervention prescribed in the protocol. The investigator or the subject may then decide to follow the protocol with less intensity. At all times during the conduct of a trial, the subject's welfare must come first and meeting those needs may not allow some aspects of the protocol to be followed. This is true for primary or secondary prevention trials, and for acute or chronic disease trials regardless of the disease. Planners of clinical trials must recognize this phenomenon and attempt to account for it in their design. Examples of adjusting for noncompliance can be found in several clinical trials.[25-32]

In the intervention group a subject who does not comply with the intervention schedule is often referred to as a "dropout." Subjects who stop the intervention regimen lose whatever potential benefit the intervention might offer. Similarly, a subject on the control regimen may at some time begin to use the intervention that is being evaluated. This subject is referred to as a "dropin." In the case of a dropin a physician may decide, for example, that surgery is required for a subject assigned to medical treatment in a clinical trial of surgery versus medical care.[29] Dropin subjects from the control group who start the intervention regimen will receive whatever potential benefit or harm that intervention might offer. Therefore, both the dropout and dropin subjects are a problem because they tend to dilute any difference between the two groups which might be produced by the intervention. This simple model does not take into account the situation in which one level of an intervention is compared to another level of the intervention. More complicated models can be developed. Regardless of the model, it must be emphasized that the assumed event rates in the control and intervention groups are modified by subjects who do not comply with the study protocol.

People who do not comply should remain in the assigned groups and be included in the analysis. The rationale for this is discussed in Chapter 16. The basic point to be made here is that eliminating subjects from analysis or transferring subjects to the other group could easily bias the results of the study. However, the observed δ is likely to be less than projected because of noncompliance and thus have an impact on the power of the clinical trial. A reduced δ, of course, means that either the sample size must be increased or the study will have smaller power than intended. Lachin[2] has proposed a simple formula to adjust crudely the sample size for a dropout rate of proportion R. The unadjusted sample size N should be multiplied by the factor $1/(1-R)^2$. Thus, if R = .25, the originally calculated sample should be multiplied by 16/9 or increased by 78%. This formula gives some quantitative idea of the effect of dropout on the sample size.

However, more refined models to adjust sample sizes for dropouts from the intervention to the control[33,34] and for dropins from the control to the intervention regimen[35] have been developed. These models adjust for the resulting changes in p_I and p_C, the adjusted rates being denoted p_I^* and p_C^*. These models also allow for another important factor, which is the time required or the intervention to achieve maximum effectiveness. For example, an antiarrhythmic drug may have an immediate effect; conversely, even though a cholesterol-lowering drug reduces serum levels quickly, it may require years to produce a maximum effect on coronary mortality.

A drug trial[27] in post myocardial infarction patients illustrates the effect of dropouts and dropins on sample size. In this trial, total mortality over a three-year follow-up period was the primary response variable. The mortality rate in the control group was estimated to be 18% (p_c = 0.18) and the intervention was believed to have the potential for reducing p_C by 28% (k = 0.28) yielding p_I = 0.1296. These estimates of p_C and k were derived from previous studies. Those studies also indicated that the dropout rate might be as high as 26% over the three years; 12% in the first year, an additional 8% in the second year, and an additional 6% in the third year. For the control group, the dropin rate was estimated to be 7% each year for a total dropin rate of 21%.

Using these models for adjustment, $p_C^* = 0.1746$ and $p_I^* = 0.1375$. Therefore, instead of $\delta = 0.0504$ (0.18–0.1296), $\delta^* = 0.0371$ (0.1746–0.1375). For a two-sided test with $\alpha = 0.05$ and $1-\beta = 0.90$, the adjusted sample size was 4020 subjects compared to an unadjusted sample size of 2160 subjects. The adjusted sample size almost doubled in this example due to the expected dropout and dropin experience and the recommended policy of keeping subjects in the originally assigned study groups. The remarkable increases in sample size because of dropouts and dropins strongly argue for major efforts to keep noncompliance to a minimum during trials.

SAMPLE SIZE CALCULATIONS FOR CONTINUOUS RESPONSE VARIABLES

For a clinical trial with continuous response variables, the previous discussion is conceptually relevant, but not directly applicable, to actual calculations. "Continuous" variables such as length of hospitalization, blood pressure, spirometric measures, neuropsychological scores and level of a serum component may be evaluated. Distributions of such measurements frequently can be approximated by a normal distribution. When this is not the case, a transformation of values, such as taking their logarithm, can still make the normality assumption approximately correct.

Suppose the primary response variable, denoted as x, is continuous with N_I and N_C subjects randomized to the intervention and control groups respectively. Assume that the variable x has a normal distribution with mean μ and variance σ^2. The true levels of μ_I and μ_C for the intervention and control groups are not known, but it is assumed that σ^2 is known. (In actual practice, σ^2 is not known and must be estimated from some data. If the data set used is reasonably large, the estimate of σ^2 can be used in place of the true σ^2. If the estimate for σ^2 is based on a small set of data, it is necessary to be cautious in the interpretation of the sample size calculations.)

The null hypothesis is $H_0: \delta = \mu_C - \mu_I = 0$ and the two-sided alternative hypothesis is $H_A: \delta = \mu_C - \mu_I \neq 0$. If the variance is known, the test statistic is:

$$Z = (\bar{x}_C - \bar{x}_I)/\sigma \sqrt{1/N_C + 1/N_I}$$

This statistic has approximately a standard normal distribution where $\bar{x}_I$ and $\bar{x}_C$ represent mean levels observed in the intervention and control groups respectively. The hypothesis-testing concepts previously discussed apply to the above statistic. If $|Z| > Z_\alpha$, then an investigator would reject H_0 at the α level of significance. By use of the above test statistic it can be determined how large a total sample $2N$ would be needed to detect a true difference δ between μ_I and μ_C with power $(1-\beta)$ and significance level α by the formula:

$$2N = \frac{4 (Z_\alpha + Z_\beta)^2 \sigma^2}{\delta^2}$$

For example, suppose an investigator wishes to estimate the sample size necessary to detect a 10 mg/dl difference in cholesterol level in a diet intervention group compared to the control group. The variance from other data is estimated to be (50 mg/dl)². For a two-sided 5% significance level, $Z_\alpha = 1.96$ and for 90% power, $Z_\beta = 1.282$. Substituting these values into the above formula, $2N = 4(1.96 + 1.282)^2(50)^2/10^2$ or approximately 1050 subjects. As δ decreases the value of $2N$ increases, and as σ^2 increases the value of $2N$ increases. This means that the smaller the difference in intervention effect an investigator is interested in detecting and the larger the variance, the larger the study must be. As with the dichotomous case, setting a smaller α and larger $1-\beta$ also increases the sample size. Figure 7-3 shows total sample size $2N$ as a function of δ/σ. As in the example, if $\delta = 10$ and $\sigma = 50$, then $\delta/\sigma = 0.2$ and the sample size $2N$ for $1-\beta = 0.9$ is approximately 1050.

Instead of looking at the difference between mean levels in the groups, an investigator interested in mean levels of change might want to test whether diet intervention lowers cholesterol from baseline levels when compared to a control. This is essentially the same question as asked before, but each subject's initial cholesterol is taken into account. Because of the likelihood of reduced variability, this type of design can lead to a smaller sample size if the question is correctly posed. Assume that Δ_C and Δ_I represent the true but unknown levels of change from baseline to some later point in the trial for the control and intervention groups, respectively. Estimates of Δ_C and Δ_I would be $\bar{x}_{C_1} - \bar{x}_{C_2}$ and $\bar{x}_{I_1} - \bar{x}_{I_2}$. These represent the differences in mean levels of the response variable at two points for each group. The investigator tests $H_0: \Delta_C - \Delta_I = 0$ versus $H_A: \Delta_C - \Delta_I = \delta \neq 0$. The variance σ_Δ^2 in this case reflects the variability of the change, and is likely to be smaller than the variability at a single measurement. This is the case if the correlation between the first and second measurements is greater than 0.5. Using δ and σ_Δ^2 as defined in this manner, the previous sample size formula and graph are applicable.

Assume that an investigator is still interested in detecting a 10 mg/dl difference in cholesterol between the two groups, but that the variance of the change is now (20 mg/dl)². The question being asked in terms of δ is approximately the same, because randomization should produce baseline mean levels in each group which are almost equal. The comparison of differences in change essentially is a comparison of the difference in mean levels of cholesterol at the second measurement. Using Figure 7-3, where $\delta/\sigma_\Delta = 10/20 = 0.5$, the sample size is 170. This impressive reduction in sample size from 1050 is due to a reduction in the variance from $(50 \text{ mg/dl})^2$ to $(20 \text{ mg/dl})^2$.

Sample Size when the Rate of Change Is the Response Variable

The previous section briefly presented the sample size calculation for trials where only two points, say a baseline and a final visit, are used to determine effect of intervention and these two points are the same for all study subjects. Often, a continuous response variable is measured at each follow-up visit. Considering only the first and last values would give one estimate of change but would not take advantage of all the available

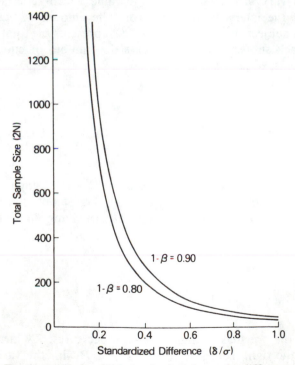

Figure 7-3 Total sample size (2N) required to detect the difference (δ) between control group mean and intervention group mean as a function of the standardized difference (δ/σ) where σ is the common standard deviation, with two-sided significance level of 0.05 and power (1-β) of 0.80 and 0.90.

data. In such a case, one possible approach is to assume that the change in response variable is approximately a linear function of time, so that the rate of change can be summarized by the slope. This model is fit to each subject's data by the standard least squares method and the estimated slope is used to summarize the subject's experience. In planning such a study, the investigator must be concerned about the frequency of the measurement and the duration of the observation period. As discussed by Schlesselman,[36] the observed measurement x can be expressed as $x = a + bt + error$, where a = intercept, b = slope, t = time, and $error$ represents the deviation of the observed measurement from a regression line. This error may be due to measurement variability, biological variability or the nonlinearity of the true underlying relationship. On the average, this error is expected to be equally distributed around 0 and have a variability denoted as σ^2 $(error)$. Schlesselman assumes that σ^2 $(error)$ is approximately the same for each subject.

The investigator evaluates intervention effectiveness by comparing the average slope in one group with the average slope in another group. Obviously, subjects in a group will not have the same slope, but the slopes will vary around some average value which reflects the effectiveness of the intervention or control. The amount of variability of slopes over subjects is denoted as σ_b^2. If D represents the total time duration for each subject and P represents the number of equally spaced measurements, σ_b^2 can be expressed as

$$\sigma_b^2 = \sigma_B^2 + \left\{ \frac{12(P\text{-}1)\sigma^2(error)}{D^2 P(P+1)} \right\}$$

where σ_B^2 is the component of subject variance in slope not due to measurement error and lack of a linear fit. The sample size required to detect difference δ between the average rates of change in the two groups is given by

$$2N = \frac{4\,(Z_\alpha + Z_\beta)^2}{\delta^2} \left[\sigma_B^2 + \frac{12(P\text{-}1)\sigma^2(error)}{D^2 P(P+1)} \right]$$

As in the previous formulas, when δ decreases, $2N$ increases. The factor on the right-hand side relates D and P with the variance components σ_B^2 and σ^2 $(error)$. Obviously as σ_B^2 and σ^2 $(error)$ increase, the total sample size increases. By increasing P and D, however, the investigator can decrease the contribution made by σ^2 $(error)$. The exact choices of P

and D will depend on how long the investigator can feasibly follow subjects, how many times he can afford to have subjects visit a clinic and other features. By manipulating P and D, an investigator can design a study which will be the most cost effective for his specific situation.

In planning for a trial, it may be assumed that a response variable declines at the rate of 80 units/year in the control group. Suppose a 25% reduction is anticipated in the intervention group. That is, the rate of change in the intervention group would be 60 units/year. Other studies provided an estimate for σ *(error)* of 150 units. Also, suppose data from a study of subjects followed every three months for one year ($D = 1$) and ($P = 5$) gave a value for the standard deviation of the slopes, $\sigma_b = 200$. The calculated value of σ_B is then 63 units. Thus, for a 5% significance level and 90% power *($Z_\alpha = 1.96$ and $Z_\beta = 1.282$)*, the total sample size would be approximately 630 for a three-year study with four visits per year ($D = 3$, $P = 13$). Increasing the follow-up time to four years, again with four measurements per year, would decrease the variability with a resulting sample size calculation of approximately 510. This reduction in sample size could be used to decide whether or not to plan a four-year study or a three-year study.

SAMPLE SIZE CALCULATIONS FOR "TIME TO FAILURE"

For many clinical trials, the primary response variable is the occurrence of some event and thus the proportion of events in each group may be compared. In these cases, the sample size methods described earlier will be appropriate. In other trials, the time to the event may be of special interest. For example, if the time to death can be increased, the intervention may be useful even though at some point the proportion of deaths in each group are similar. Methods for analysis of this type of outcome are generally referred to as lifetable or survival analysis methods. (See Chapter 14.)

The basic approach is to compare the survival curves for the groups. A survival curve may be thought of as a graph of the probability of surviving up to any given point in time. The methods of analysis now widely used are non-parametric; that is, no mathematical model about the shape of the survival curve is assumed. However, for the purpose of estimating sample size, some assumptions are often made. A common model assumes that the survival curve, $S(t)$, follows an exponential distribution, $S(t) = e^{-\lambda t}$, where λ is called the hazard rate or force of mortality. Using this model, survival curves are characterized by λ. Thus, the survival curves from a control and an intervention group can be compared by testing H_0: $\lambda_C = \lambda_I$. An estimate of λ is obtained as the inverse of the mean survival time. Sample size formulations have been considered by several investigators.[37-39] One simple formula is given by

$$2N = \frac{4 (Z_\alpha + Z_\beta)^2}{[\ln(\lambda_C/\lambda_I)]^2}$$

where N is the size of the sample in each group and Z_α and Z_β are defined as before. As an example, suppose one assumes that the force of mortality is .30 in the control group and .20 for the intervention being tested; that is, $\lambda_C/\lambda_I = 1.5$. If $\alpha = .05$ (two-sided) and $1-\beta = .90$, then $N = 128$ or $2N = 256$. The corresponding mortality rates for five years of follow-up are .7769 and .6321 respectively. Using the comparison of two proportions, the total sample size would be 412. Thus, the time to failure method may give a more efficient design, requiring a smaller number of subjects.

The method just described assumes that all patients will be followed to the event. With few exceptions, clinical trials with a survival outcome are terminated at time T before all subjects have died. For the patients still alive, the time to death is said to be censored at time T. For this situation, Lachin[2] gives the approximate formula

$$2N = \frac{2(Z_\alpha + Z_\beta)^2 [\phi(\lambda_C) + \phi(\lambda_I)]}{(\lambda_I - \lambda_C)^2}$$

where $\phi(\lambda) = \lambda^2/(1 - e^{-\lambda T})$ and where $\phi(\lambda_C)$ or $\phi(\lambda_I)$ are defined by replacing λ with λ_C or λ_I, respectively. If a five year study were being planned ($T = 5$) with the same design specifications as above, $2N = 376$. Thus, the loss of information due to censoring must be compensated for by increasing the sample size.

If the subjects are to be recruited continually during the five years of the trial, the formula given by Lachin is identical but with $\phi(\lambda) = \lambda^3 T/(\lambda T - 1 + e^{-\lambda T})$. Using the same design assumptions, $2N = 620$, showing that not having all the subjects at the start requires an additional increase in sample size.

More typically subjects are recruited uniformly over a period of time, T_0, with the trial continuing for T years ($T > T_0$). In this situation, the sample size can be estimated as before using

$$\phi(\lambda) = \frac{\lambda^2}{1 - [e^{-\lambda(T - T_0)} - e^{-\lambda T}]/\lambda T_0}$$

Here, the sample size ($2N$) of 466 is between the previous two examples suggesting that it is preferable to get subjects recruited as rapidly as possible.

Further models are given by Lachin.[2] A useful series of nomograms for sample size considering factors such as α, $1-\beta$, the subject recruitment time, the follow-up period, and the ratio of the hazard rates have been published.[40]

One of the methods used for comparing survival curves is the proportional hazards model or the Cox regression model which is discussed briefly in Chapter 14. For this method, sample size estimates have been provided.[41,42] As it turns out, the formula by Schoenfeld for the Cox model[41] is identical to that given above for the simple exponential case, although developed from a different point of view.

All of the above methods assume that the hazard rate remains constant during the course of the trial. This may not be the case. The Beta-Blocker Heart Attack Trial[27] compared three-year survival in two groups of subjects with intervention starting one to three weeks after a myocardial infarction. The risk of death was high initially, decreased steadily, and then became relatively constant.

For cases where the event rate is relatively small and the clinical trial will have considerable censoring, most of the statistical information will be in the number of events. Thus, the sample size estimates using simple proportions will be quite adequate. In the Beta-Blocker Heart Attack Trial, the three year control group event rate was assumed to be 0.18. For the intervention group, the event rate was assumed to be approximately 0.13. In the situation of $\phi(\lambda) = \lambda^2(1 - e^{-\lambda T})$, a sample size $2N = 2208$ is obtained. In contrast, the unadjusted sample size using simple proportions is 2160. Again, it should be emphasized that all of these methods are only approximations and the estimates should be viewed as such.

SAMPLE SIZE FOR TESTING "EQUIVALENCY" OF INTERVENTIONS

In some instances, an effective intervention has already been established and is considered the standard. New interventions under consideration may be less expensive, have fewer side effects, or have less impact on an individual's general quality of life, and thus may be preferred. This issue is common in the pharmaceutical industry where a product developed by one company may be tested against an established intervention manufactured by another company. Studies of this type are sometimes referred to as trials with positive controls.

Given that several trials have shown that certain beta-blockers are effective in reducing mortality in post-myocardial infarction patients,[27,43,44] it is likely that any new beta-blockers developed will be tested against proven agents. The Nocturnal Oxygen Therapy Trial[45] tested whether the daily amount of oxygen administered to chronic obstructive lung disease patients could be reduced from 24 hours to 12 hours without impairing oxygenation. The Intermittent Positive Pressure Breathing[28] trial considered whether a simple and less expensive method for delivering a bronchodilator into the lungs would be as effective as a more expensive device. A cancer trial compared the tumor regression rates between subjects receiving the standard, diethylstilbestrol, or the newer agent, tamoxifen.[46] As no significant difference in effectiveness was found, tamoxifen was recommended because of less toxicity.

The problem in designing positive control studies is that there can be no statistical method to demonstrate complete equivalence. Failure to reject the null hypothesis is not sufficient reason to claim two interventions to be equal but merely that the evidence is inadequate to say they are different.[47] Computing a sample size assuming no difference, using the previously described formulae, results in an infinite sample size.

While demonstrating complete equivalence is an impossible task, one possible approach has been discussed.[48,49] The strategy is to specify some value, δ, such that interventions with differences which are less than this might be considered "equally effective" or "equivalent." Specification of δ may be difficult but without it, no study can be designed. The methods developed require that if the two interventions really are equally effective, the upper $100(1-\alpha)\%$ confidence interval for the intervention difference will not exceed δ with the probability of $1-\beta$. One can alternatively approach this from a hypothesis testing point of view, stating the null hypothesis that the two interventions differ by less than δ.

For studies with a dichotomous response, one might assume the event rate for the two interventions to be equal to p (ie, $p = p_C = p_I$). This leads to a sample size formula of

$$2N = 4p(1-p)(Z_\alpha + Z_\beta)^2/\delta^2$$

where N, Z_α and Z_β are defined as before. This formula differs slightly from the analogue presented earlier due to the different way the hypothesis is stated. The formula for continuous variables,

$$2N = 4\sigma^2(Z_\alpha + Z_\beta)^2/\delta^2$$

turns out to be identical to its analogue. Blackwelder and Chang[50] give graphical methods for computing sample size estimates for studies of equivalency.

Another proposed strategy for comparing a new to a standard drug is to show bioequivalence or similarity in bioavailability. Several authors have discussed this approach.[51-54] If two formulations are within specified limits for a profile of biochemical measurements and one of them has already been proven to be effective, the argument is made that further efficacy trials are not necessary. The sample size estimation for demonstrating bioequivalence poses the same problem as described above and the approach is similar.

MULTIPLE RESPONSE VARIABLES

This book has stressed the advantages of having a single primary question and a single primary response variable, but clinical trials occasionally have more than one of each. More than one question may be asked because investigators cannot agree about which factor is most important. As an example, one clinical trial involving two schedules of oxygen administration to patients with chronic obstructive lung disease had three major questions in addition to comparing the mortality rate.[45] Measures of pulmonary function, neuropsychological status, and quality of life were evaluated. For the subjects, all three are important.

Sometimes more than one primary response variable is used to assess a single primary question. This may reflect uncertainty as to how the investigator can answer the question. A clinical trial involving patients with pulmonary embolism[55] employed three methods of assessing a drug's ability to resolve emboli. They were: lung scanning, arteriography, and hemodynamic studies. Another trial involves the use of drugs to limit myocardial infarct size.[56] Precordial electrocardiogram mapping, radionuclide studies, and enzyme levels all are used to evaluate the effectiveness of the drugs.

Computing a sample size for such clinical trials is not easy. One could attempt to define a single model for the multidimensional response and use one of the previously discussed formulas. Such a method would require several assumptions about the model and its parameters and might require information about correlations between different measurements. Such information is rarely available. A more reasonable procedure would be to compute sample sizes for each individual response variable. If the results give about the same sample size for all variables, then the issue is resolved. However, more commonly, a range of sample sizes will be obtained. The most conservative strategy would be to use the largest sample size computed. The other response variables would then have even greater power to detect the hoped-for reductions or differences (since they required smaller sample sizes). Unfortunately, this approach is the most expensive and difficult to undertake. Of course, one could also choose the smallest sample size of those computed. That would probably not be desirable, because the other response variables would have less power than usually required, or only larger differences than expected would be detectable. It is possible to select a middle range sample size, but there is no assurance that this will be appropriate. An alternative approach is to look at the difference between the largest and smallest sample sizes. If this difference is very large, the assumptions that went into the calculations should be re-examined and an effort should be made to resolve the difference.

As will be discussed in Chapter 16, when multiple comparisons are made, the chance of finding a significant difference in one of the comparisons (when, in fact, no real differences exist between the groups) is greater than the stated significance level. In order to maintain an appropriate significance level α for the entire study, the significance level required for each test to reject H_0 should be adjusted.[16] The significance level required for rejection (α') in a single test can be approximated by α/k where k is the number of multiple response variables. For several response variables this can make α' fairly small (eg, $k = 5$ implies $\alpha' = 0.01$ for each of k response variables with an overall $\alpha = 0.05$). If the correlation between response varibles is known, then the adjustment can be made more precisely.[57] In all cases, the sample size would be much larger than if the use of multiple response variables were ignored, so that most studies have not strictly adhered to this solution of modifying the significance level. Some investigators, however, have attempted to be conservative in the analysis of results.[58] There is a reasonable limit as to how much α' can be decreased in order to give protection against false rejection of the null hypothesis. Some investigators have chosen $\alpha' = 0.01$ regardless of the number of tests. In the end, there are no easy solutions. A somewhat conservative value of α' needs to be set and the investigator needs to be aware of the multiple testing problem during the analysis.

REFERENCES

1. Freiman, J.A. Chalmers, T.C., Smith, H., Jr. et al. The importance of beta, the type II error and sample size in the design and interpretation of the randomized control trial: survey of 71 "negative" trials. *N Engl J Med.* 299:690–694, 1978.

2. Lachin, J.M. Introduction to sample size determination and power analysis for clinical trials. *Controlled Clin Trials* 2:93–113, 1981.

3. Brown, B.W., Jr. Statistical controversies in the design of clinical trials — some personal views. *Controlled Clin Trials* 1:13–27, 1980.

4. Altman, D.G. Statistics and ethics in medical research: III. How large a sample? *Br Med J.* 281:1336–1338, 1980.

5. Gore, S.M. Assessing clinical trials — trial size. *Br Med J.* 282:1687–1689, 1981.

6. Fleiss, J.L. *Statistical Methods for Rates and Proportions.* 2nd ed. New York: John Wiley and Sons, 1981.

7. Snedecor, G.W., and Cochran, W.G. *Statistical Methods.* 6th ed. Ames: Iowa State University Press, 1967.

8. Brown, B.W., and Hollander, M. *Statistics: A Biomedical Introduction.* New York: John Wiley and Sons, 1977.

9. Remington, R.D., and Schork, M.A. *Statistics With Applications to the*

Biological and Health Sciences. Englewood Cliffs, N.J.: Prentice-Hall, 1970.

10. Dixon, W.J., and Massey, F.J. Jr. *Introduction to Statistical Analysis*. 3rd Ed. New York: McGraw-Hill, 1969.

11. Armitage, P. *Statistical Methods in Medical Research*. New York: John Wiley and Sons, 1971.

12. Schlesselman, J.J. Planning a longitudinal study: I. Sample size determination. *J Chronic Dis*. 26:553-560, 1973.

13. Brittain, E., and Schlesselman, J.J. Optimal allocation for the comparison of proportions. *Biometrics* 38:1003-1009, 1982.

14. Rothman, K.J. A show of confidence. *N Engl J Med*. 299:1362-1363, 1978.

15. McHugh, R.B., and Le, C.T. Confidence estimation and the size of a clinical trial. *Controlled Clin Trials* 5:157-163, 1984.

16. Armitage, P., McPherson, C.K., and Rowe, B.C. Repeated significance tests on accumulating data. *J R Stat Soc.*, Series A 132:235-244, 1969.

17. Gail, M., and Gart, J.J. The determination of sample sizes for use with the exact conditional test in 2×2 comparative trials. *Biometrics* 29:441-448, 1973.

18. Gail, M. The determination of sample sizes for trials involving several independent 2×2 tables. *J Chronic Dis*. 26:669-673, 1973.

19. Haseman, J.K. Exact sample sizes for use with the Fisher-Irwin Test for 2×2 tables. *Biometrics* 34:106-109, 1978.

20. Feigl, P. A graphical aid for determining sample size when comparing two independent proportions. *Biometrics* 34:111-122, 1978.

21. Casagrande, J.T., Pike, M.C., and Smith, P.G. An improved approximate formula for calculating sample sizes for comparing two binomial distributions. *Biometrics* 34:483-486, 1978.

22. Ury, H.K., and Fleiss, J.L. On approximate sample sizes for comparing two independent proportions with the use of Yates' Correction. *Biometrics* 36:347-351, 1980.

23. Fleiss, J.L., Tytun, A., and Ury, H.K. A simple approximation for calculating sample sizes for comparing independent proportions. *Biometrics* 36:343-346, 1980.

24. Wacholder, S., and Weinberg, C.R. Paired versus two-sample design for a clinical trial of treatments with dichotomous outcome: power considerations. *Biometrics* 38:801-812, 1982.

25. Coronary Drug Project Research Group. The Coronary Drug Project. Design, methods, and baseline results. *Circulation* 47(Suppl I):I1-I79, 1973.

26. Aspirin Myocardial Infarction Study Research Group. A randomized controlled trial of aspirin in persons recovered from myocardial infarction. *JAMA* 243:661-669, 1980.

27. Beta-Blocker Heart Attack Trial Research Group. A randomized trial of propranolol in patients with acute myocardial infarction. I. Mortality results. *JAMA* 247:1707-1714, 1982.

28. The Intermittent Positive Pressure Breathing Trial Group. Intermittent positive pressure breathing therapy of chronic obstructive pulmonary disease — a clinical trial. *Ann Intern Med*. 99:612-620, 1983.

29. CASS Principal Investigators and Their Associates. Coronary Artery Surgery Study (CASS): A randomized trial of coronary artery bypass surgery: survival data. *Circulation* 68:939-950, 1983.

30. Hypertension Detection and Follow-up Program Cooperative Group. Five-year findings of the Hypertension Detection and Follow-up Program. Reduction in mortality of persons with high blood pressure, including mild hypertension. *JAMA* 242:2562-2571, 1979.

31. Multiple Risk Factor Intervention Trial Research Group. Multiple Risk Factor Intervention Trial. Risk factor changes and mortality results. *JAMA* 248:1465–1477, 1982.

32. Collaborative Group on Antenatal Steroid Therapy. Effect of antenatal dexamethasone administration on the prevention of respiratory distress syndrome. *Am J Obstet Gynecol.* 141:276–287, 1981.

33. Halperin, M., Rogot, E., Gurian, J., and Ederer, F. Sample sizes for medical trials with special reference to long-term therapy. *J Chronic Dis.* 21:13–24, 1968.

34. Schork, M.A., and Remington, R.D. The determination of sample size in treatment-control comparisons for chronic disease studies in which drop-out or non-adherence is a problem. *J Chronic Dis.* 20:233–239, 1967.

35. Wu, M., Fisher, M., and DeMets, D. Sample sizes for long-term medical trials with time-dependent dropout and event rates. *Controlled Clin Trials* l:109–121, 1980.

36. Schlesselman, J.J. Planning a longitudinal study: II. Frequency of measurement and study duration. *J Chronic Dis.* 26:561–570, 1973.

37. Pasternack, B.S., and Gilbert, H.S. Planning the duration of long-term survival time studies designed for accrual by cohorts. *J Chronic Dis.* 24:681–700, 1971.

38. Pasternack, B.S. Sample sizes for clinical trials designed for patient accrual by cohorts. *J Chronic Dis.* 25:673–681, 1972.

39. George, S.L., and Desu, M.M. Planning the size and duration of a clinical trial studying the time to some critical event. *J Chronic Dis.* 27:15–24, 1974.

40. Schoenfeld, D.A., and Richter, J.R. Nomograms for calculating the number of patients needed for a clinical trial with survival as an endpoint. *Biometrics* 38:163–170, 1982.

41. Schoenfeld, D.A. Sample-size formula for the proportional-hazards regression model. *Biometrics* 39:499–503, 1983.

42. Freedman, L.S. Tables of the number of patients required in clinical trials using the logrank test. *Stat Med.* 1:121–129, 1982.

43. Hjalmarson, A., Herlitz, J., Malek, I., et al. Effect on mortality of metoprolol in acute myocardial infarction: a double-blind randomized trial. *Lancet* ii:823–827, 1981.

44. The Norwegian Multicenter Study Group. Timolol-induced reduction in mortality and reinfarction in patients surviving acute myocardial infarction. *N Engl J Med.* 304:801–807, 1981.

45. Nocturnal Oxygen Therapy Trial Group. Continuous or nocturnal oxygen therapy in hypoxemic chronic obstructive lung disease: a clinical trial. *Annals Int Med.* 93:391–398, 1980.

46. Ingle, J.N., Ahmann, D.L., Green, S.J., et al. Randomized clinical trial of diethylstilbestrol versus tamoxifen in postmenopausal women with advanced breast cancer. *N Engl J Med.* 304:16–21, 1981.

47. Spriet, A., and Beiler, D. When can "nonsignificantly different" treatments be considered as "equivalent"? (letter to the editors) *Br J Clin Pharmacol Ther.* 7:623–624, 1979.

48. Makuch, R., and Simon, R. Sample size requirements for evaluating a conservative therapy. *Cancer Treat Rep.* 62:1037–1040, 1978.

49. Blackwelder, W.C. "Proving the null hypothesis" in clinical trials. *Controlled Clin Trials* 3:345–353, 1982.

50. Blackwelder, W.C. and Chang, M.A. Sample size graphs for "proving the null hypothesis." *Controlled Clin Trials* 5:97–105, 1984.

51. Anderson, S., and Hauck, W.W. A new procedure for testing equivalence in comparative bioavailability and other clinical trials. *Commun Statist-Theor Meth*. 12:2663–2692, 1983.

52. Dunnett, C.W., and Gent, M. Significance testing to establish equivalence between treatments, with special reference to data in the form of 2×2 tables. *Biometrics* 33:593–602, 1977.

53. Westlake, W.J. Statistical aspects of comparative bioavailability trials. *Biometrics* 35:273–280, 1979.

54. Westlake, W.J. Response to "Bioequivalence testing—a need to rethink." *Biometrics* 37:591–593, 1981.

55. Urokinase Pulmonary Embolism Trial Study Group. Urokinase-streptokinase embolism trial: phase II results. *JAMA* 229:1606–1613, 1974.

56. Roberts, R., Croft, C., Gold, H.K., et al. Effect of propranolol on myocardial infarct size in a randomized blinded multicenter trial. *N Engl J Med*. 311:218–225, 1984.

57. Miller, R.G. *Simultaneous Statistical Inference*. New York: McGraw-Hill, 1966.

58. Coronary Drug Project Research Group. Clofibrate and niacin in coronary heart disease. *JAMA* 231:360–381, 1975.

CHAPTER 8

Baseline Assessment

In clinical trials, baseline refers to the status of a subject before the start of intervention. Baseline data may be measured by interview, questionnaire, physical examination, and laboratory tests. Measurement need not be only numerical in nature. It can also mean classification of study participants into categories based on factors such as absence or presence of some trait or condition.

As discussed in Chapter 3, baseline data describe the people studied, enabling the scientific community to compare the trial results with those of other studies. Different results from different studies may be attributed to seemingly minor differences in the study subjects. These differences, in turn, can lead to the creation of new hypotheses, which may be tested.

A number of clinical trials of aspirin and other platelet-active drugs have been conducted in post-myocardial infarction patients. The trials differed with respect to time to enrollment after the infarction, as well as in dosage. There were also apparent differences in outcome, which could have been due to time of initiation of intervention. Studies that enrolled subjects early after the infarction tended to have more favorable trends than those recruited later. In addition, in some trials, the subgroups of subjects enrolled earlier seemed to benefit more from the intervention than those entered later.[1] These observations led to the start of the Persantine–Aspirin Re-Infarction Study II, which limited enrollment to the early recovery phase after an infarction.

Valid inferences about benefits of therapy depend on the kinds of people enrolled, as well as on study design. Complete reporting of baseline data allows clinicians to evaluate a new therapy's chance of success in their patients. On the other hand, judgment should be used in determining the factors to be measured at baseline. Evaluating factors which are unlikely to be pertinent to the trial is not only wasteful of money and time but may reduce subject cooperation. This chapter is concerned with the uses of baseline data, what constitutes a baseline measure, and assessment of baseline comparability.

FUNDAMENTAL POINT

Relevant baseline data should be measured in all study subjects before the start of intervention.

USES OF BASELINE DATA

Although baseline data may be used to determine the eligibility of subjects, it is assumed that any subjects who are found unable to meet entrance criteria have been excluded from the study before assignment to either intervention or control. The characteristics of people not enrolled are of interest when attempting to generalize the results of the trial (Chapter 3). For the discussion in this chapter, however, only data from enrolled subjects are considered.

Analysis of Baseline Comparability

Baseline data allow people to evaluate whether the study groups were comparable before intervention was started. The assessment of comparability should include risk or prognostic factors, pertinent demographic and socioeconomic characteristics, and medical history. This assessment is necessary in both randomized and non-randomized trials. While randomization on the average produces balance between comparison groups, it does not guarantee balance in any specific trial. This may even be true in large studies. In the Aspirin Myocardial Infarction Study,[2] which had over 4500 subjects, the aspirin group was at slightly higher risk than the placebo group when baseline characteristics were examined. In assessment of comparability in any trial, the investigator can look at only factors about which he is aware. Obviously, those which are unknown cannot be compared.

Stratification and Subgrouping

If there is concern that one or two key prognostic factors may not "balance out" during randomization, thus yielding noncomparable groups at baseline, the investigator may stratify on the basis of these factors. Stratification can be done at the time of randomization or during analysis. Chapters 5 and 16 review the advantages and disadvantages of stratified randomization and stratified analysis. The point here is that, in order to stratify at either time, the relevant characteristics of the subjects at baseline must be known. For non-randomized trials, these factors must also be measured in order to select properly the control group and analyze results by strata.

Often, investigators are interested not only in the response to intervention in the total study group, but also in the response in one or more subgroups. Particularly, in studies in which an overall intervention effect is present, analysis of results by appropriate subgroup may help to identify the specific population most likely to benefit from, or be harmed by, the intervention. Subgrouping may also help to elucidate the mechanism of action of the intervention. Definition of such subgroups should rely only on baseline data, not data measured after initiation of intervention (except for factors such as age or sex which cannot be altered by the intervention). An example of establishing subgroups is the Canadian Cooperative Study Group trial of aspirin and sulfinpyrazone in people with cerebral or retinal ischemic attacks.[3] After noting an overall benefit from aspirin in reducing continued ischemic attacks or stroke, the authors observed that the benefit was restricted to men. Any conclusions drawn from subgroup hypotheses not explicitly stated in the protocol, however, should be given much less credibility than those from hypotheses stated *a priori*. Retrospective subgroup analysis should serve primarily to generate new hypotheses for subsequent testing (Chapter 16).

Prognostic Variables

Baseline measurements enable investigators to perform natural history analyses in a control group which is on either placebo or no uniformly administered intervention. The prognostic importance of suspected risk factors for a variety of fatal and nonfatal events can be evaluated, particularly in large, long-term trials. This evaluation can include verification of previously ascertained risk factors as well as identification of others not earlier considered. Such analyses, although peripheral to the main objectives of a clinical trial, may be important for future research efforts. Their potential importance is especially true if variables which are subject to intervention can be identified.[4] Even if they are not variables which can be studied in future trials, they can be used in future stratification or subgroup analyses.

Response Variables and Toxicity

Making use of baseline data will usually add sensitivity to a study. For example, an investigator may want to evaluate a new antihypertensive agent. He can either compare the mean change in blood pressure from baseline to some subsequent time in the intervention group against the mean change in the control group, or simply compare the mean blood pressures of the two groups at the end of the study. The former method usually is a more powerful statistical technique because it can reduce the variability of the response variables. As a consequence, it may permit either fewer subjects to be studied or a smaller difference between groups to be detected. This technique was used in a study of thrombolytic agents in the treatment of pulmonary embolism.[5]

Evaluation of possible unwanted reactions requires knowledge—or at least tentative ideas—about what effects might occur. The investigator should record at baseline those clinical or laboratory features which are likely to be adversely affected by the intervention. Unexpected adverse reactions might be missed, but the hope is that animal studies or earlier clinical work will have identified the important factors to be measured.

WHAT IS A TRUE BASELINE MEASUREMENT?

In order to describe accurately the study subjects, baseline data should ideally reflect the true condition of the subjects. Certain information can be obtained accurately by means of one measurement or evaluation at a baseline interview and examination. However, for many variables, accurately determining the subject's true state is difficult, since the mere fact of impending enrollment in a trial or the baseline examination itself may alter a measurement. For example, is true blood pressure reflected by a single measurement taken at baseline? If more than one measurement is made, which one should be used as the baseline value? Is the average of repeated measurements recorded over some extended period of time more appropriate? Does the subject need to be taken off all medications or be free of other factors which might affect the determination of a true baseline level? When resolving these questions, the screening required to identify eligible subjects, the time and cost entailed in this identification, and the specific uses for the baseline information must be taken into account.

In almost every clinical trial, some sort of screening of potential subjects is necessary. Screening eliminates subjects who, based on the entrance criteria, are ineligible for the study. Enrollment of a large number of ineligible subjects reflects on the quality of the trial and may cause problems in data analysis. Occasionally, entrance criteria require repeated measurements over a period of time. A prerequisite for inclusion is the subject's willingness to comply with a possibly long and arduous study protocol. The subject's commitment, coupled with the need for repeated measurements of eligibility criteria, means that intervention allocation usually occurs later than the time of the investigator's first contact with the subject. A problem may result from the fact that discussing a study with someone or inviting him to participate in a clinical trial may alter his state of health. For instance, a person asked to join a study of lipid-lowering agents because he had an elevated serum cholesterol at a screening examination might change his diet on his own initiative just because of the fact he was invited to join the study. Therefore, his serum cholesterol as determined at baseline, perhaps a month after the initial screen, may be somewhat lower than usual. Improvement could happen in many potential candidates for the trial and could affect the validity of the assumptions used to calculate sample size. If the study calls for a special dietary regimen, this might not be so effective at the new, lowered cholesterol level. As a result of the modification in subject behavior, there may be less room for change due to the intervention.

Although it may be impossible to avoid altering the behavior of potential subjects, in study design it is often possible to adjust for such changes. Special care can be taken when discussing studies with people to avoid sensitizing them. Time between invitation to join a study and baseline evaluation can be kept to a minimum. People who have greatly changed their eating habits between the initial screen and baseline, as determined by a questionnaire at baseline, can be declared ineligible to join. Alternatively, they can be enrolled and the required sample size increased. Whatever is done, these are expensive ways to compensate for the reduced expected effectiveness of the intervention.

Sometimes a person's eligibility for a study is determined by measuring continuous variables, such as blood pressure or cholesterol level. If the entrance criterion is a high or low value, a phenomenon referred to as "regression toward the mean" is encountered.[6] Regression toward the mean occurs because measurable characteristics of an individual do not have constant values but vary above and below the average value for that individual. Because of this variability, although the population mean for a characteristic may be relatively constant over time, the locations of individuals within the population change. If two sets of measurements are made on individuals within the population, therefore, the correlation between the first and second series of measurements will not be perfect. That is, depending on the variability, the correlation will be something less than 1. In addition, it is often the case that the more distant a measured characteristic is from the population mean of that characteristic, the more variable the measurement tends to be.

Therefore, whenever subjects are selected from a population on the basis of some measured characteristic, the mean of a subsequent measurement will be closer to the population mean than is the first measurement mean. Furthermore, the more extreme the initial selection criterion (that is, the further from the population mean), the greater will be the regression toward the mean at the time of the next measurement.

The "floor-and-ceiling effect" used as an illustration by Schor[7] is helpful in understanding this concept. If all the flies in a closed room are near the ceiling in the morning, than at any subsequent time during the day more flies will be below where they started than above. Similarly, if the flies start close to the floor, the more probable it is for them to be higher, rather than lower, at any subsequent time.

Cutter[8] gives some non-biological examples of regression toward the mean. He presents the case of a series of three successive tosses of two dice. The average of the first two tosses is compared with the average of the second and third tosses. If no selection or cut-off criterion is used, the average of the first two tosses would, in the long run, be close to the average of the second and third tosses. However, if a cut-off point is selected which restricts the third toss to only those instances where the average of the first and second tosses is nine or greater, regression toward the mean will occur. The average of the second and third tosses for this selected group will be less than the average of the first two tosses for this group.

As with the example of the subject changing his diet between screening and baseline, this phenomenon of regression toward the mean can complicate assessment of intervention. In another case, an investigator may wish to evaluate the effects of an antihypertensive agent. He measures blood pressure once at the baseline examination and enters into his study only those people with diastolic pressures over 95 mm Hg. He then gives a drug and finds when he rechecks that most people have responded with lowered blood pressures. However, when he re-examines the control group, he finds that most of those people also have lower pressures. The value of a control group is obvious in such situations. An investigator cannot simply compare pre-intervention and post-intervention values in the intervention group. He must compare post-intervention values in the intervention group with values obtained at similar times in the control group. This regression toward the mean phenomenon can also lead to a problem discussed previously. People are screened initially for high or low values. Because of regression, the values at baseline are less extreme than the investigator had planned on, and there is less room for improvement from the intervention. In the blood pressure example, after randomization, many of the patients may have diastolic blood pressures in the low 90's or even 80's rather than above 95 mm Hg. There may be a reluctance to use antihypertensive agents in people with pressures this low and, certainly, the opportunity to demonstrate full effectiveness of the agent may be lost.

Two approaches to reducing the impact of regression toward the mean were used by the Hypertension Detection and Follow-up Program.[9] One approach was to use a more extreme value than the entrance criterion when the investigators screened people before baseline. Secondly, in order to enroll subjects with diastolic blood pressure greater than 90 mm Hg, each potential subject had his pressure recorded three times. Only those whose second and third measure averaged 95 mm Hg or greater were invited to the clinic for further evaluation. The average of two recordings at the second evaluation was the baseline value, which was used for comparison with subsequent determinations.

When baseline data are measured too far in advance of intervention assignment, a response may occur in the interim. An investigator might not be able to determine whether this response occurred before or after the beginning of intervention. An effective treatment may reduce the frequency of complications of a disease from 15% to 10% in three months. The investigator should be able to detect this improvement. If, however, responses occur in the one-month interval between baseline examination and start of intervention at a rate of 5% per month in each group, he would see a reduction from nearly 20% to nearly 15%. This reduction of about 25%, instead of 33%, might not be statistically significant, and the investigator could arrive at a different conclusion. The subjects having events in the interval between allocation and the actual initiation of intervention would dilute the results and decrease the chances of finding a significant difference. In the European Coronary Surgery Study[10] coronary artery bypass surgery should have taken place within three months of intervention allocation. However, the mean time until surgery was 3.9 months. Consequently, of the 21 deaths in the surgical group, six occurred before surgery could be performed. If the response, such as death, is non-recurring and this occurs between baseline and the start of intervention, the number of subjects at risk of having the event later is reduced. Therefore, the investigator needs to be alert to any event occurring after baseline but before intervention is instituted. When such an event occurs before allocation to intervention or control, he can exclude the subject from the study. When the event occurs after allocation, but before start of intervention, subjects should nevertheless be kept in the study and the event counted in the analysis. Removal of such subjects from the study may bias the outcome. For this reason, the European Coronary Surgery Study Group kept such subjects in the trial for purposes of analysis. In an unblinded trial of alprenolol in survivors of an acute myocardial infarction,[11] a large number of subjects were excluded between randomization and start of therapy. Since the study was not blinded, bias could have entered into the decision to exclude subjects. Bias seems all the more likely since the investigators excluded many more patients from the alprenolol group than from the control group. The appropriateness of withdrawing subjects from data analysis is discussed more fully in Chapter 16.

Particularly troublesome are those studies where baseline factors cannot be completely ascertained until after intervention has begun. The

Multi-Institutional Study to Limit Infarct Size[12] investigated whether the size of a myocardial infarct could be reduced by propranolol or hyaluronidase administered within hours of onset of the infarction. Unfortunately, because of the requirement to start intervention early, the investigators could not fully characterize the infarct in the time available. Determination of the initial extent of the infarct became extraordinarily difficult, if not impossible, afterward—especially since the interventions were postulated to alter its course. In this instance, the investigators needed to be satisfied with more limited baseline information than they would have liked.

Even if an investigator can get baseline information just before initiating intervention, he may need to compromise. For instance, being an important prognostic factor, serum cholesterol level is obtained in most studies of heart disease. Serum cholesterol levels, however, are temporarily lowered during the acute phase of a myocardial infarction. Therefore, in any trial using people who have just had a myocardial infarction, baseline serum cholesterol data relate poorly to their usual levels. Only if the investigator has data on subjects from a time before the myocardial infarction would usual cholesterol levels be known. Cholesterol levels obtained at the time of the infarction might not allow him to evaluate natural history or make reasonable observations about changes in cholesterol that occurred because of the intervention. On the other hand, because he has no reason to expect that one group would have greater lowering of cholesterol at baseline than the other group, such levels can certainly tell him whether the study groups are initially comparable.

Medications that subjects are taking may also complicate the interpretation of the baseline data and restrict the uses to which an investigator can put baseline data. Hospitalized patients with severe cardiac arrhythmias will likely be given short-acting antiarrhythmic drugs. The investigator may want to evaluate the efficacy of a new drug in the long-term suppression of cardiac arrhythmias. However, for the purpose of obtaining a baseline estimate of the arrhythmias, to discontinue the present medication may be unethical. This issue came up in the Beta-blocker Heart Attack Trial.[13] Although the primary response variable was mortality, the effect of propranolol on cardiac arrhythmias was obviously of interest. However, at baseline, a number of subjects were being prescribed antiarrhythmic agents such as lidocaine. These could not be withdrawn simply to obtain an uncontaminated estimate of baseline arrhythmia. Some baseline data, therefore, became the number of subjects on lidocaine and other antiarrhythmic drugs, rather than the frequency of severe arrhythmias.

Appreciating that, for many measurements, baseline data may not reflect the subject's true condition at the time of baseline, investigators perform the examination as close to the time of intervention allocation as possible. Baseline assessment may, in fact, occur shortly after allocation but prior to the actual start of intervention. The advantage of such timing is that the investigator does not spend extra time and money performing baseline tests on subjects who may turn out to be ineligible. The baseline examination then occurs immediately after randomization and is performed not to exclude participants, but solely as a baseline reference point. Since allocation has already occurred, all subjects remain in the trial regardless of the findings at baseline. This reversal of the usual order is not recommended in single-blind or unblinded studies, because it raises the possibility of bias during the examination. If the investigator knows to which group the subject belongs, he may subconsciously measure characteristics differently, depending on the group assignment. Furthermore, the order reversal may unnecessarily prolong the interval between intervention allocation and its actual start.

BALANCE AND IMBALANCE

As mentioned earlier, assessment of baseline comparability is one of the reasons for measuring baseline variables. Imbalance in important characteristics can yield misleading results. Assessment of baseline comparability is important in all trials, and particularly so in non-randomized studies. The investigator needs to look at baseline variables in several ways. The simplest is to compare each variable to make sure that it has reasonably similar distribution in each study group. Means, medians, and ranges are all convenient measures. The investigator can also combine the variables, giving each one an appropriate weight or coefficient, but doing this presupposes a knowledge of the relative prognostic importance of the variables. This kind of knowledge can come only from another study with a very similar population or by looking at the control group after the present study is completed. The weighting technique has the advantage that it can take into account numerous small differences between groups. If imbalances between most of the variables are in the same direction, the overall imbalance can turn out to be large, even though differences in individual variables are small.

In small studies especially, randomization may not yield entirely comparable groups. Treatment and control groups may differ in one or more baseline variables. In the 30-center Aspirin Myocardial Infarction Study which involved over 4500 subjects,[2] each center can be thought of as a small study with about 150 subjects. When the baseline comparability within each center was reviewed, substantial differences in almost half the centers were found, some favoring intervention and some, control (Furberg, CD, unpublished data). The difference between intervention and control groups in predicted three-year mortality, using the Coronary

Drug Project model[14] exceeded 20% in five of the 30 clinics. Therefore, all factors which are known or suspected to be important in the subsequent course of the condition under study should be looked at when interpreting results. Identified imbalances do not invalidate a randomized trial, but they may make interpretation of results more complicated. In non-randomized trials, the credibility of the findings may be compromised because selection bias may have occurred.

When comparing baseline factors, remember that groups can never be shown to be identical. Only absence of "significant" differences can be demonstrated. In fact, 5% of the comparisons would be expected to show differences at the 0.05 significance level. In studies with small sample sizes, clinically important differences might be ignored because there is insufficient power to demonstrate statistically significant differences. Yet in numerous reports, authors state that the groups are similar because no statistically significant differences have been detected. Such a statement is an unwarranted interpretation of the data.

Of course, all important prognostic factors have probably not been identified—nor can all of them be measured. For example, ability to predict coronary heart disease occurrence has increased impressively in recent years.[15] Nevertheless, there are still large gaps in knowledge. This is one of the reasons for randomization, and why the process of identifying imbalances in known risk factors may not give a complete, or even true, picture of baseline comparability.

REFERENCES

1. Canner, P.L. Aspirin in coronary heart disease: comparison of six clinical trials. *Isr J Med Sci*. 19:413–423, 1983.

2. Aspirin Myocardial Infarction Study Research Group. A randomized, controlled trial of aspirin in persons recovered from myocardial infarction. *JAMA* 243:661–669, 1980.

3. The Canadian Cooperative Study Group. A randomized trial of aspirin and sulfinpyrazone in threatened stroke. *N Engl J Med*. 299:53–59, 1978.

4. Coronary Drug Project Research Group. Treatable risk factors—hypercholesterolemia, smoking, and hypertension—after myocardial infarction: implications of the Coronary Drug Project data for clinical management. *Primary Care* 7:175–179, 1980.

5. Urokinase Pulmonary Embolism Trial Group. Urokinase Pulmonary Embolism Trial: phase 1 results. *JAMA* 214:2163–2172, 1970.

6. James, K.E. Regression toward the mean in uncontrolled clinical studies. *Biometrics* 29:121–130, 1973.

7. Schor, S.S. The floor-and-ceiling effect. *JAMA* 207:120, 1969.

8. Cutter, G.R. Some examples for teaching regression toward the mean from a sampling viewpoint. *Am Stat*. 30:194–197, 1976.

9. Hypertension Detection and Follow-up Program Cooperative Group. Five-year findings of the Hypertension Detection and Follow-up Program. I.

Reduction in mortality of persons with high blood pressure, including mild hypertension. *JAMA* 242:2562–2571, 1979.

10. European Coronary Surgery Study Group. Coronary-artery bypass surgery in stable angina pectoris: survival at two years. *Lancet* 1:889–893, 1979.

11. Ahlmark, G., and Saetre, H. Long-term treatment with β-blockers after myocardial infarction. *Eur J Clin Pharmacol.* 10:77–83, 1976.

12. Roberts, R., Croft, C., Gold, H.K., et al. Effect of propranolol on myocardial-infarct size in a randomized blinded multicenter trial. *N Engl J Med.* 311:218–225, 1984.

13. Beta-Blocker Heart Attack Trial: Protocol. National Heart, Lung, and Blood Institute, Division of Heart and Vascular Diseases, Clinical Trials Branch. Bethesda, MD., 1978.

14. Coronary Drug Project Research Group. Factors influencing long-term prognosis after recovery from myocardial infarction—three year findings of the Coronary Drug Project. *J Chronic Dis.* 27:267–285, 1974.

15. Gordon, T., Kannel, W.B., and Halperin, M. Predictability of coronary heart disease. *J Chronic Dis.* 32:427–440, 1979.

CHAPTER 9

Recruitment of Study Subjects

Often the most difficult task in a clinical trial involves obtaining sufficient study subjects within a reasonable time. Time is a factor for both scientific and logistic reasons. From a scientific point of view, the clinical trial should be completed while the primary question is still of interest. Of course, a scientific question likely to become obsolete in the near future probably does not merit a clinical trial. In terms of logistics, the longer recruitment goes on beyond the initially allotted time the more planning is disrupted, costs increase, and personnel changes are likely to occur. The primary reasons for recruitment failure are inadequate planning and overoptimism. Invariably, subject recruitment is more difficult than originally planned.

Recruitment of subjects will vary depending on the kind and size of the study, the length of time available, the setting (hospital, clinic, workplace), whether the study is single or multicenter, and numerous other factors. Therefore, this book will not elaborate on specific techniques. Rather, concepts and general methods will be summarized, with emphasis on anticipating and preventing problems.

FUNDAMENTAL POINT

In evaluating the feasibility of recruiting study subjects, investigators need to BE CONSERVATIVE, to establish interim recruitment goals, and to have contingency plans available for immediate implementation.

PLANNING STAGE

In the planning stage of a trial, an investigator needs to evaluate realistically the likelihood of obtaining sufficient study participants within the allotted time. He should obtain the best estimate of how many subjects who meet the specific entrance criteria are available to him. When reviewing census tract data or hospital and physician records, he

121

must remember that such data may be out of date, incomplete, or incorrect. People may have moved or died since the records were last updated. Information about current use of drugs, or frequency of surgical procedures may not reflect what will be the case in the future, when the trial is actually conducted. Records may not give sufficient—or even accurate—details about potential subjects and will certainly not contain information about willingness of people to participate initially or about their degree of cooperation.

After initial record review, an investigator may find he needs to expand his population base by increasing his geographical catchment area, by canvassing additional hospitals, by relaxing one or more of the study entrance criteria, by increasing the planned recruitment time or by a combination of these factors. The initial survey of patient sources should be as thorough as possible, because these determinations are better made before a study starts than afterward.

Planning also involves making sure that the investigator has the necessary organizational assistance. The degree of support for his efforts from his institution and colleagues should be ascertained. There may be other people in the institution or nearby institutions competing for similar subjects. Since it is not advisable to have a subject in more than one trial at a time, competing studies could decrease the likelihood that the investigator will make his recruitment goal. If the recruitment goal cannot be accomplished, a reappraisal regarding entrance criteria, study time frame—or even the feasibility of conducting the study at all—is needed.

Overoptimistic recruitment projections are the rule rather than the exception. In a trial of aspirin in people with transient ischemic attacks,[1] estimates of numbers of potential subjects were made by each collaborating investigator. According to the study's Manual of Instruction, approximately 300 subjects would be enrolled annually. Instead, it took almost three years of recruitment to achieve this number. Clinics in the Coronary Primary Prevention Trial[2] had a similar experience. A recruitment effort that was expected to take one year took two and a half years. The problems encountered and the solutions developed in that trial, especially with regard to occupational and community screening and the use of mass mailing and media, have been extensively described.[3] Baines has reported on difficulties in recruiting subjects for the Canadian National Breast Screening Study.[4] After a comprehensive program of educating the medical profession and the public, and obtaining media support, enrollment improved markedly. The best approach involved mailing personally addressed letters to potentially eligible subjects, followed by telephone calls. These examples are not unusual. In such circumstances, studies either extend recruitment, accept fewer subjects, or, commonly, do both. In accepting a smaller number of subjects than hoped for, the investigator must either alter design features, such as the primary response variable, or change assumptions about intervention effectiveness and subject compliance. As indicated previously, such changes midway in a trial may be liable to legitimate criticism.

In the examples given, recruitment difficulties were able to be identified because the studies made initial sample size calculations. Unfortunately, many studies either do not make such calculations or do not publish them.

If the data concerning recruitment of potential trial subjects is truly scanty, a pilot or feasibility study might be undertaken. Information on optimal sources of subjects, recruitment techniques, and estimates of yields can be obtained.

In a trial of elderly people the question arose whether people in their 70s or 80s would be willing to volunteer and actively participate in a long-term placebo-control trial. Therefore, before implementing a costly full-scale trial, a pilot study was conducted to answer these and other questions.[5] The pilot study showed that the elderly were willing participants and provided important information on recruitment techniques. As a result of the success of the pilot, a full-scale trial was started.

CONDUCT OF RECRUITMENT

Successful recruitment of subjects depends not only on planning but also on the commitment of the investigators. Part of planning is the need to have a system both for identifying all potential subjects from the recruitment pool and for screening these people to see if they meet eligibility criteria. For hospital-based studies, logging all admissions to special units, wards, or clinics is invaluable. However, keeping such records complete can be difficult, especially during evenings or weekends. During such hours, those most dedicated to the study are often not available to ensure accuracy and completeness. In addition, during vacation times and illness, there may be no one to keep the log up to date. Therefore, frequent quality checks should be made. Subject privacy is also important. Do investigators have a right to keep a record on people who have not consented to be listed? The answer to this will vary from institution to institution and will depend on who is keeping the log and for what reason. If information can be recorded by using code numbers only, that can often help maintain privacy.

For population-based recruitment efforts, investigators need to identify times when they are likely to reach the maximum number of potential subjects. If they intend to make visits to homes or hope to contact people by means of telephone calls, they should count on working evenings or weekends. Unless potential subjects are retired, or investigators plan on contacting people at their jobs (which, depending on the nature of the job, may be difficult), normal working hours may not be productive times. Vacation periods and summers are also often slow periods for recruitment.

For the success of most studies it is essential to establish short-term and long-term recruitment goals. The investigator should record these goals and make every effort to achieve them. Since lagging recruitment commonly results from a slow start, timely preparation and establishing initial goals are crucial. The investigator should be ready to begin recruiting subjects on the first official day of the study.

In a long-term study, the use of interim goals helps keep the investigator from falling too far behind and may avoid a grossly uneven recruitment pace. Inasmuch as subject follow-up is usually done at regular intervals, uneven recruitment results in periods of peak and slack during the follow-up phase. This threatens effective use of staff time and equipment. Of course, establishing a goal in itself does not guarantee keeping subject recruitment on target. The goals need to be realistic and the investigator must commit himself to meeting each interim goal. Formal statistical models have been proposed for determining and assessing interim recruitment goals.[6]

If some centers in a multicenter study fall behind goal, an attempt should be made to discover the reasons as soon as possible. Often, valuable insight can be obtained by comparing results from different centers. Techniques can be learned from the centers performing best, and the others should be encouraged to incorporate these techniques. Multicenter studies require a central office to oversee recruitment, to compare results, and to lend support and encouragement. Both scientific and financial rewards can be incorporated into such studies. Frequent feedback to the centers by means of tables and graphs, which show how the actual recruitment compares with the originally projected goals, are useful tools. Examples are shown in the following figures and table. Figure 9-1 shows the progress of an investigator who started subject recruitment on schedule and maintained his pace during the recruitment period. He accurately assessed his sources of subjects and was committed to enrolling subjects in a relatively even fashion. Figure 9-2 shows the record of an investigator who started slowly. After a while, he began to improve. However, considerable effort was required for him to compensate for the poor start. His efforts included expanding the base from which his subjects were recruited and increasing the time spent in enrollment. In contrast, as seen in Figure 9-3, the investigator started slowly and was never able to improve his performance. This center was dropped from a multicenter study because it could not contribute enough subjects to the study to make its continued participation efficient. Table 9-1 shows goals, actual recruitment and projected final totals (assuming no change in enrollment pattern) for three other centers of a multicenter trial.

Such tables are useful in looking at study recruitment at a given time and in projecting final numbers of subjects. They should be updated as often as necessary.

The actual mechanics of recruiting subjects need to be worked out in advance. Keeping potential subjects waiting to be interviewed and evaluted is a poor way to earn their confidence. Investigators must be certain that necessary staff, facilities, and equipment are available at the proper times in the proper places. If many potential subjects are being

Table 9-1
Weekly Recruitment Status Report by Center*

Center	(1) Contracted Goal	(2) Enrollment This Week	(3) Actual Enrollment to Date	(4) Goal Enrollment to Date	(5) Actual Minus Goal	(6) Success Rate (3)/(4)	(7) Final Projected Intake	(8) Final Deficit or Excess (7)-(1)
A	150	1	50	53.4	− 3.4	0.94	140	− 10
B	135	1	37	48.0	− 11.0	0.77	104	− 31
C	150	2	56	53.4	2.6	1.05	157	7

*Table used in the Beta-blocker Heart Attack Trial: Coordinating Center, University of Texas, Houston.

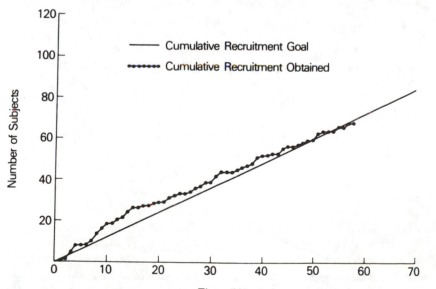

Figure 9-1 Subject recruitment in a clinic that consistently performed at goal rate.

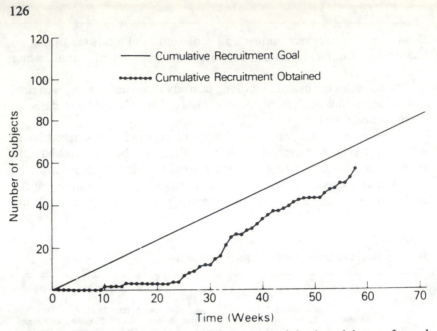

Figure 9-2 Subject recruitment in a clinic that started slowly and then performed at greater than goal rate.

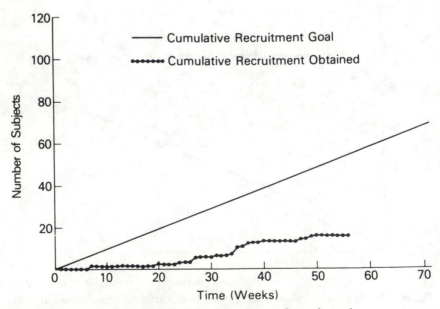

Figure 9-3 Subject recruitment in a clinic that performed poorly.

seen, laying out timetables and flow charts may ensure that screening proceeds smoothly. Such charts should include the number of people to be seen at each step in the process at a given time, the number and type of personnel and amount of time required to process each subject at each step, and the amount of equipment needed (with an allowance for "down" time). A planned pilot phase is helpful in making these assessments. One positive aspect of slow early recruitment is that the "bugs" in the start-up process can be worked out and necessary modifications made. The logistics of the recruitment effort are not always simple when follow-up of previously enrolled subjects occurs while investigators are still recruiting. In long-term studies, the most difficult time is usually toward the end of the recruitment phase when many subjects are in various stages of recruitment and follow-up. This is true because typically the same staff, space and equipment are used for recruitment, screening, and baseline and for follow-up examinations.

A contingency plan should be available in case recruitment lags. Additional sources of potential study subjects should be kept in reserve. Approval from hospital staff, corporation directors or others who control large numbers of potential subjects often takes considerable time. Waiting until recruitment problems are identified before initiating such approval can lead to weeks or months of inaction. Therefore, making plans to go to these other sources before the study actually gets underway is advised. If they are not needed, little is lost except some additional time used in planning. Most of the time these reserves will be used.

Finally, the investigator should monitor recruitment status at regular and frequent intervals. Review of these data with his staff keeps everyone aware of how recruitment is progressing. If recruitment lags, the delay can be noted early, the reasons identified and appropriate action taken.

All of these aspects require careful advance planning and clear delineation of staff responsibilities. They also require a person in charge who will make sure they get carried out. However, even the best planning is not enough. Commitment and willingness to spend a considerable amount of time in the recruitment effort are equally important. Just as investigators usually overestimate the number of subjects available, they underestimate the time and effort needed to recruit. Investigators must accommodate themselves to the schedules of potential subjects, many of whom work. Thus, recruitment is often done on weekends and evenings, as well as during usual working hours.

GENERAL TECHNIQUES

As mentioned above, the details of specific recruitment techniques will not be discussed because they vary with the particular trial. Rather, this section will summarize some basic strategies of subject recruitment. First, the investigator actively searches for subjects. He identifies target groups such as hospitalized persons, clinic patients, patients of particular physicians, census tract populations, or employees of factories or other organizations. He then screens these groups, either by searching records or setting up a testing facility for rapidly processing potential subjects (eg, special laboratory tests in clinic patients or blood pressure determinations in a group of factory workers). Those people passing the initial screen or chart review are invited to have a more detailed eligibility evaluation. This method can be time-consuming and tedious but is often more fruitful than other methods.

In one study of cholesterol reduction in men with heart disease, the investigator had an elaborate, well-organized scheme. He identified all hospitals in a specific geographic area. Letters were mailed to these hospitals asking for the names and addresses of all patients who had a heart attack within the past few years. When information such as sex, age, and attending physician was available, that too was requested. When there was no response within a certain time, further letters were sent to the hospital. After getting the names of the patients, the investigator contacted the attending physicians asking for permission to speak with, or write to, their patients. If approval was received, the patients were contacted to see whether they were interested in possibly entering the study. Unless these people were obviously ineligible for enrollment, they were invited to a screening visit. Some hospitals refused (on grounds of confidentiality) to send out names of patients. In these instances, the investigator asked the hospital to forward letters prepared by him to the attending physicians and patients. Finally, each attending physician was kept aware of the status of his patient. The steps described above comprise a long, arduous process, resulting in a low yield. However, for certain trials the scheme may be the most effective way of ensuring that all potential subjects are identified.

Another strategy involves directly inviting study subjects. Unlike the first method, no attempt is made to pre-screen the people who are contacted about the trial. Because of that, it may be more appealing and require less initial work. Subject solicitation may be done through mass media, through wide dissemination of leaflets which advertise the trial, or through participation by the investigator in health fairs or similar vehicles. None of these methods is "foolproof." The yield is unpredictable and seems to depend predominantly on the skill with which the approach is made and the size and kind of audience it reaches. One success story featured a distinguished investigator in a large city who managed to get on a local television station's early evening news show. Thousands of people volunteered for the screening program following this single five-minute appeal.

Often a combined approach is fruitful. In one large study, the investigator identified possible subjects and wrote letters to them, inviting them to participate. He received a poor response until he publicized his study via local radio and television news and interview programs. The media coverage had somehow "legitimized" the study, as well as having reminded people about it.

Subjects may also be approached through a third party. For example, an investigator may bring the attention of physicians to his study by means of letters, telephone calls, talks at professional society meetings, notices in professional journals or exhibits at scientific conferences. The hope is that these physicians will identify a potential subject and either notify the investigator or ask the person to call him. However, usually this technique results in very few eligible subjects. In one multicenter trial an investigator invited internists and cardiologists from a large metropolitan area to a meeting. He described the study, its importance and his need to recruit men who had had a myocardial infarction. Each of the physicians stood up and promised to contribute one or more subjects. One hundred fifty subjects were pledged; five were finally referred. Despite this, such pleas may be worthwhile because they make the professional community aware of a study and its purposes. Announcements should be made before the study begins recruiting subjects. The courtesy of informing area professionals about the study in advance helps in enlisting cooperation or, at least, in reducing opposition. If the investigator is recruiting ill subjects, physicians will then not be surprised by hearing of the study from their patients before learning about it from the investigator. When use of this technique is contemplated, professional talks or notices should indicate whether the investigator is simply notifying physicians of his study or whether he is actively seeking their support in recruiting subjects.

Whichever of the three methods is used, several points are worth making:

1. Success of a technique is unpredictable. What works in one place at one time may not work at the same place at another time — or the converse. Therefore, the investigator needs to be flexible and to leave room for modifications.
2. If the investigator is working with subjects who are ill, he must maintain good relationships with their personal physicians. If physicians don't approve of the study or of the way it is being conducted, they will urge their patients not to participate.
3. Investigators must respect the families of potential subjects. Most subjects like to discuss participation in experiments with their family and friends. Investigators

should be prepared to review thoroughly the study with these people. If the study requires long-term cooperation from the subject, we encourage such discussions. Anything that increases family support is likely to be reflected in better recruitment and protocol compliance.

4. Recruiting should not be too aggressive. While encouragement is necessary, excessive efforts to convince people to participate could be harmful in the long run. People reluctant to join may be more likely to abandon the study later.

RECRUITMENT PROBLEMS

Even when carefully planned and perfectly executed, recruitment may still proceed slowly. Investigators should count on "Murphy's Law" operating and expect problems to occur despite their efforts. These may be completely unforeseen. In one multicenter study, there were reports of murders of inpatients at the hospital adjacent to the study clinic. It is hardly surprising that attendance at the clinic fell off sharply.

One possible way of dealing with lagging recruitment is to accept a smaller number of subjects than originally planned. Doing this is far from ideal, inasmuch as the power of the study will be reduced. Only if the investigator is lucky and discovers that some of the assumptions used in calculating sample size were too pessimistic would this "solution" be acceptable. The previously mentioned trial of aspirin in people with transient ischemic attacks[1] had the happy experience of finding that aspirin produced a greater effect than initially postulated. Therefore, the less-than-hoped-for number of subjects turned out to be adequate. Alternatively, an extra effort might be made to achieve better-than-projected subject compliance to study protocol, thereby reducing the number of required subjects.

A second approach is to relax the inclusion criteria. Relaxing inclusion criteria should be done only if there is little expectation that the study design will suffer. The design can be marred when, as a result of the new type of subjects, the control group event rate is altered to such an extent that the estimated sample size is no longer appropriate. Also, the expected response to intervention in the new subjects may not be as great as in the original subjects. Furthermore, the intervention might have a greater likelihood of being harmful in the new subjects than in those originally recruited. Finally, the addition of different subjects at a late stage could lead to imbalance in baseline characteristics. The difference in additional subjects would not matter if the proportion of subjects randomized to each group stayed the same throughout recruitment. However, as indicated in Chapter 5, certain randomization schemes alter that proportion, depending on baseline criteria or study results. Under these circumstances, changing entrance criteria can create imbalances.

In the Coronary Drug Project,[7] only people with documented transmural myocardial infarctions were originally eligible. With enrollment falling behind, the investigators decided to admit subjects with non-transmural myocardial infarctions. Inasmuch as there was no reason to expect that the action of lipid-lowering agents would be any different in the new group than in the original group and since the lipid-lowering agents were not contraindicated in the new subjects, the modification seemed reasonable. However, there was some concern that the control group mortality experience would be changed because mortality in people with non-transmural infarctions could be less than mortality in people with transmural infarctions. Nevertheless, the pressure of recruitment overrode that concern. Possible baseline imbalance did not turn out to be a problem. The percentages of people with non-transmural infarctions in each of the treatment groups were quite close (16.7, 18.3, 16.4, 18.6, 19.4, and 16.7 in the five treatment groups and the placebo group respectively). The overall percent of such people was 17.5. In this particular study, where the total number of subjects was so large (8341), there was every expectation that randomization would yield comparable groups. If there had been uncertainty regarding this, stratified randomization could have been employed (Chapter 5). Including people with non-transmural infarctions may have reduced the power of the study because they had a lower mortality rate than those with transmural infarctions in each of the treatment groups, including the placebo group. However, the treatments were equally ineffective when people with transmural infarctions were looked at separately from people with non-transmural infarctions.[8]

A possible third solution to recruitment problems is to extend the time for recruitment or, in the case of multicenter studies, to add recruiting centers. Both are the preferred solutions, requiring neither modification of admission criteria nor diminution of power. However, they are also the most costly. Whether the solution of additional time or additional centers is adopted depends on cost, on the logistics of finding and training new high quality centers, and on the need for obtaining study results quickly.

It has been claimed that one of the factors leading to recruitment problems is unwillingness on the part of the people and their private physicians to participate in a randomized trial. Zelen[9] proposed a design where eligible patients were randomized to control or intervention before informed consent is obtained. Only those assigned to the intervention are informed of the trial and asked to give their consent. The other group receives standard treatment and are followed by their usual source of medical care.

This technique was employed in a trial of different surgical approaches for breast cancer.[10] After 44 months, only 519 subjects had been enrolled, 16% of the goal. The major reasons given by the investigators were concern that randomization would impair the doctor-patient relationship, difficulty in obtaining informed consent, and difficulty separating the scientist and clinician roles. After switching to the Zelen design, recruitment increased six fold.[11] In other trials, the experience has been encouraging but not as dramatic. However, several concerns have been raised.[11-13] These include the ethics of not informing the subjects assigned to the control group that they are in a trial, and the potential dilution of results if a large number of subjects decline to participate. In fact, the Cancer Research Council Working Party in Great Britain has rejected the Zelen model.[13] The long-term acceptability and utility of this approach to improving recruitment remains unclear.

Collins et al[14] have reviewed ways in which investigators handled lagging recruitment in seven Veterans Administration multicenter trials. In six of the studies, the required sample size was reevaluated and lowered. The authors indicate that the recruitment problems could have been avoided had the planning of the trial been better.

When recruitment becomes difficult, a sensitive issue is the possibility that an investigator will loosely interpret entrance criteria or will deliberately change data in order to enroll otherwise ineligible subjects. Unfortunately, this issue is not merely theoretical. Such practices have occurred, to a very limited extent, in more than one trial. The best way to avoid the problem is to make it clear that, if this occurs, both the study and the subjects can be harmed and that neither science nor the investigators are served by such practices. Random record audits by an independent person or group may also serve as a deterrent.

RECYCLING OF POTENTIAL SUBJECTS

In most studies, investigators struggle to find a sufficient number of subjects. When a prospective subject just misses meeting the eligibility criteria, the temptation is natural to try to enroll him by repeating a measurement, perhaps under slightly different conditions. In general, this "recycling" should be discouraged. A study is harmed by enrolling persons for whom the intervention might not be effective. Most important, the subject possibly will take an intervention for which he has no indication and which might be detrimental to him.

Instances exist where, in order to enter a drug study, the subject needs to be off all other medication with similar actions. At baseline, he may be asked whether he has complied with this requirement. If he has not, the investigator may repeat the instructions and have the subject return in a week for repeat baseline measurements. This "second chance" is different from recycling. The entrance criterion checks on a subject's ability to comply with a protocol and his understanding of instructions. From a design point of view, it may be legitimate to give a candidate this second chance. However, the second-chance subject, even if he passes the repeat baseline measurement, may not be as good a candidate for the study as someone who complied on the first occasion.[15] Therefore, exclusion of these subjects is recommended.

REFERENCES

1. Fields, W.S., Lemak, N.A. Frankowski, R.F. et al. Controlled trial of aspirin in cerebral ischemia. *Stroke* 8:301–316, 1977.

2. Agras, W.S., and Marshall, G. Recruitment for the Coronary Primary Prevention Trial. *Clin Pharmacol Ther.* 25:688–690, 1979.

3. Agras, W.S., Bradford, R.H., and Marshall, G.D., (Eds): Recruitment for clinical trials: the Lipid Research Clinics Coronary Primary Prevention Trial experience. Its implications for future trials. *Circulation* 66(Suppl IV):IV-1–IV-78, 1982.

4. Baines, C.J. Impediments to recruitment in the Canadian National Breast Screening Study: response and resolution. *Controlled Clin Trials* 5:129–140, 1984.

5. Hughes, G.H., and Schnaper, H.W. The Systolic Hypertension in the Elderly Program. *Int J Ment Health* 11:76–97, 1982.

6. Lee, Y.J. Interim recruitment goals in clinical trials. *J Chron Dis.* 36:379–389, 1983.

7. Coronary Drug Project Research Group. The Coronary Drug Project: design, methods, and baseline results. *Circulation* 47 (suppl 1):1–50, 1973.

8. Coronary Drug Project Research Group. Clofibrate and niacin in coronary heart disease. *JAMA* 231:360–381, 1975.

9. Zelen, M. A new design for randomized clinical trials. *N Engl J Med.* 300:1242–1245, 1979.

10. Taylor, K.M., Margolese, R.G., and Soskolne, C.L. Physicians' reasons for not entering eligible patients in a randomized clinical trial of surgery for breast cancer. *N Engl J Med.* 310:1363–1367, 1984.

11. Ellenberg, S.S. Randomization designs in comparative clinical trials. *N Engl J Med.* 310:1404–1408, 1984.

12. Angell, M. Patients' preferences in randomized clinical trials (editorial). *N Engl J Med.* 310:1385–1387, 1984.

13. Consent: how informed? (editorial). *Lancet* i:1445–1447, 1984.

14. Collins, J.F., Bingham, S.F., Weiss, D.G., et al. Some adaptive strategies for inadequate sample acquisition in Veterans Administration cooperative clinical trials. *Controlled Clin Trials* 1:227–248, 1980.

15. Sackett, D.L. A compliance practicum for the busy practitioner. (Eds): R.B. Haynes, D.W. Taylor, D.L. Sackett. In *Compliance in Health Care.* Baltimore: Johns Hopkins University Press, 1979.

CHAPTER 10

Data Collection and Quality Control

No study is better than the quality of its data. In clinical trials data are collated from several sources — interviews, questionnaires, subject examinations or laboratory determinations. Also, data that have been collected and evaluated by someone outside the study can be used in a trial; for example, diagnoses obtained from death certificates or hospital records.

Quality control starts with clear definitions of response variables and procedures and with training; it ends with data collection, editing and assessment. Results from any experiment, either based on poorly standardized procedures that use ambiguous definitions or conducted by insufficiently trained staff with limited knowledge about the study protocol, can lead to erroneous results and conclusions. This chapter will review why problems in data collection arise, and provide some general solutions. The section on quality monitoring emphasizes issues that need to be considered in drug trials.

FUNDAMENTAL POINT

During all phases of a study, sufficient effort should be spent to ensure that all key data are of high quality.

PROBLEMS IN DATA COLLECTION

Problems in data collection can be of several sorts and can apply to the initial acquisition of data (eg, physical examination) as well as to the recording of the data on a form or data entry into a remote computer terminal or microcomputer. First, incomplete and irretrievable data can haunt the investigator. Such a situation arises, for example, from the inability of subjects to provide necessary information, from inadequate physical examinations, from laboratory mishaps or from carelessness in study form completion or data entry. The percent of missing data in a study can be considered as one indicator of the quality of the data and, therefore, the quality of the trial.

Second, erroneous data may not be recognized and, therefore, can be even more troublesome than incomplete data. For study purposes, a specified condition may be defined in a particular manner. A clinic staff member may unwittingly use a clinically acceptable definition, but one that is different from the study definition. Specimens may be mislabeled. In one clinical trial, the investigators suspected a mislabeling when, in a glucose tolerance test, the fasting glucose levels were higher than the one-hour glucose levels in some subjects. Badly calibrated equipment can be a source of error. In addition, the wrong data may be entered on a form. A blood pressure of 84/142 mm Hg, rather than 142/84 mm Hg, is easy to identify as wrong. However, while 124/84 mm Hg may be incorrect, it is perfectly reasonable, and the error would not necessarily be recognized.

The third problem is variability in the observed characteristics. Variability reduces the opportunity to detect real changes. The variability between repeated assessments can be unsystematic (or random), systematic, or a combination of both. Variability can be intrinsic to the characteristic being measured, the instrument used for the measurement, or the observer responsible for obtaining the data. The problem of variability is not unique to any specific field of investigation.[1,2] Reports of studies of repeat chemical determinations, determinations of blood pressure, physical examinations, and interpretations of x-rays, electrocardiograms and histological slides[3-14] indicate the difficulty in obtaining reproducible data. People perform tasks differently and may vary in knowledge and experience. These factors can lead to inter-observer variability. In addition, inconsistent behavior of the same observer between repeated measurements may also be much greater than expected. While less than inter-observer variability, intra-observer inconsistency nevertheless can be appreciable.

In 1947, Belk and Sunderman[3] reviewed the performance of 59 hospital laboratories on several common chemical determinations. Using prepared samples, they found that "unsatisfactory results outnumbered the satisfactory." In 1969, Lewis and Burgess[4] assessed interlaboratory measures of red blood cell count using two methods (visual and electronic). The ranges for both methods were extremely broad. Results for the visual method varied from 2.2×10^6 RBC/mm^3 to 5.1×10^6 RBC/mm^3 and for the electronic method from 0.7×10^6 RBC/mm^3 to 4.7×10^6 RBC/mm^3. In 1978, others[5] looked at six selected laboratories. The overall performance was reasonably good. However, performance on simulated patient specimens was worse than when designated quality-control specimens were analyzed. This indicates that when special attention is given to the analyses, laboratories perform better. Classification of histologic specimens can also be highly variable. Feinstein and colleagues[6] reviewed interpretation of cellular types of lung cancer by five experienced pathologists. On second readings, pathologists disagreed with their first diagnoses up to 20% of the time. When Davies[7] looked at reading of electrocardiograms by nine experienced cardiologists, he found large inter-observer and intra-observer disagreement. On the average, on re-reading electrocardiograms, the cardiologists disagreed with one in eight of their original interpretations. A study that assessed clinical findings of the respiratory system[8] concluded that "inter-observer repeatability of respiratory signs falls midway between chance and total agreement."

Regardless of the source of variability, several factors may have an impact on the magnitude of variability. Vagueness in definitions, inadequate methodology, lack of training of personnel and carelessness can all increase the variation inherent in any measurement.

MINIMIZING POOR QUALITY DATA

General approaches for minimizing potential problems in data collection are summarized below. Most of these should be considered during the planning phase of the trial. Examples in the cardiovascular field are provided by Rose and Blackburn.[15]

Clear definitions of entry and diagnostic criteria and methodology are essential. These should be written so that all investigators and staff can apply them in a consistent manner throughout the trial. The same question can be interpreted in many ways. Even the same investigator may forget how he previously interpreted a question unless he can readily refer to instructions and definitions. Accessibility of these definitions is also important. Ideally, the study forms should contain all necessary information. If that is not possible, the forms should outline the key information and refer the investigator to the appropriate page in a manual of procedures. Such a manual of procedures should be prepared in every clinical trial. This document provides detailed answers to all conceivable "how to" questions. Although it may contain information about study background, design, and organization, the manual of procedures is not simply an expanded protocol. In addition to listing eligibility criteria and response variable definitions, it should indicate how the criteria and variables are determined. Most important, the manual needs to describe the subject visits—their scheduling and content—in detail. Instructions for filling out forms, performing tasks such as laboratory determinations, drug ordering, storing and dispensing, and compliance monitoring must be accurate and complete. Finally, recruitment techniques, informed consent, subject safety, emergency unblinding, use of concomitant therapy and other issues need to be addressed. Updates and clarifications usually occur during the course of a study. These revisions should be made available to every staff person involved in data collection.

Descriptions of laboratory methodology and the ways the results are to be reported also need to be stated in advance. In one study, plasma levels of the drug propranolol were determined by using four methods. Only after the study ended was it discovered that two laboratories routinely were measuring free propranolol and two other laboratories were measuring propranolol hydrochloride. A conversion factor allowed investigators to make simple adjustments and arrive at legitimate comparisons. Such adjustments are not always possible.

Training sessions for investigators and staff to promote standardization of procedures are crucial to the success of any large study. Whenever more than one person is filling out forms or examining subjects, training sessions help to minimize error. There may be more than one correct way of doing something in clinical practice, but for study purposes, there is only one way. Similarly, the questions on a form should always be asked in the same way. The answer to, "Have you had any chest pain in the last three months?" may be different from the answer to, "You haven't had any chest pain in the last three months, have you?" Training laboratory personnel is equally important. Two technicians may use slightly different techniques. These differences can lead to confusing results. Kahn and colleagues[16] reviewed the impact of training procedures instituted in the Framingham Eye Study. The two days of formal training included duplicate examinations and discussions about differences, and the use of a reference set of fundus photographs.

Mechanisms to verify that all clinic staff do things the same way should be developed. These could include instituting certification procedures for specified types of data collection. If blood pressure, electrocardiograms, pulmonary function tests or laboratory tests are important, the people performing these determinations should not only be trained, but also be tested and certified as competent. Periodic retraining and certification are especially useful in long-term studies since people tend to forget, and there is likely to be personnel turnover. For situations where staff must conduct clinical interviews, special training procedures to standardize the approach have been used.[17]

Well-designed forms will minimize errors and variability. Forms should be as short and as well organized as possible, with a logical sequence to the questions. Forms should be clear, with few "write-in" answers. As little as possible should be left to the imagination of the person completing the form. This means, in general, no essay questions. The questions should elicit the necessary information and little else. Questions which are tacked on because the answers would be "nice to know" are rarely analyzed and may distract attention from pertinent questions. In several studies where death is the primary response variable, investigators have expressed interest in learning about the circumstances surrounding the death. In particular, the occurrence of symptoms before death, the time lapse from the occurrence of such symptoms until death, and the activity and location of the subject at the time of death have been considered important. While this may be true, focusing on it has led to the creation of extraordinarily complex forms which take considerable time to complete. Moreover, questions arise concerning the accuracy of the information, because much of it is obtained from sources who may not have been with the subject when he died. Unless investigators clearly understand how these data will be used, simpler forms are preferable. The experience regarding forms and recommendations from the Coronary Drug Project have been reported.[18] Wright and Haybittle have also provided general guidelines to forms design for clinical trials.[19]

Pretesting of forms and procedures is useful. Several people similar to the intended subjects should participate in a simulated interview and examination to make sure procedures are properly performed and questions on the forms flow well and provide the desired information. Furthermore, by pretesting, the investigator grows familiar and comfortable with the form. Fictional case histories can be used to check form design and the care with which forms are completed. When developing forms, most investigators cannot conceive of the numerous ways questions can be misinterpreted until several people have been given the same information and asked to fill out the same form. Part of the reason for different answers is undoubtedly due to carelessness. Misinterpretation may not be detected when forms are filled out on real subjects. Anyone editing a form has no way of identifying errors which are not completely unreasonable. Inadequacies in form structure and logic can also be uncovered by use of pretesting. In conclusion, pretesting reveals areas where forms might be improved and where additional training might be worthwhile.

There is little point in constructing fictional case histories unless there is an opportunity for follow-up discussion. This helps people completing the forms to understand how the forms are meant to be completed and what interpretations are wanted. Discussion also alerts them to carelessness. When done before the start of the study, follow-up discussion allows the investigator to modify inadequate items on forms. These case history exercises might be profitably repeated several times during the course of a long-term study to indicate when education and retraining are needed. Ideally, forms should not be changed after the study has started. Inevitably, though, modifications are made. Pretesting can help to minimize this.

Both variability and bias in the assessment of response variables should be minimized through repeat assessment, blinded assessment, or (ideally) both. At the time of the examination of a subject, for example, an investigator may determine blood pressure two or more times and record the average. Performing the measurement without knowing the group assignment helps to minimize bias. In unblinded or single-blinded studies, the examination might be performed by someone other than the investigator. For blood pressure, another method of minimizing bias might be the use of a random zero device.[13] In assessing slides, x-rays, or electrocardiograms, two individuals can make independent, blinded evaluations, and the results can be averaged or adjudicated in cases of disagreement. Independent evaluations are particularly important when the assessment requires an element of judgment. Classification of response variables such as cause of death or nonfatal events can be performed in a similar manner.

The introduction of microcomputers into clinical trials has the potential for improving data quality. Programs have been developed that identify missing, extreme or inconsistent values and which prohibit further data entry until a correction has been made. In cases where an investigator must go back to the subject to check the information, this aspect is particularly valuable, because the error is identified rapidly. Double entry of data is also used to reduce the error rate. Adequate training of staff is essential for fully realizing the advantages of this technology.

An issue being debated is whether forms can be eliminated. Typically, a paper form is completed and the data transferred to the microcomputer. Thus, a record trail is available for data verification and audit. If only the final entered data are available, there is no assurance that the data have not been altered inappropriately.

Programs can be developed which will ensure that both original and revised data are saved. Thus, a computerized audit trail can be developed. In such a case, it is conceivable that an investigator can dispense with paper forms.

In general, there has been a favorable experience with entering clinical trial data into microcomputers. Error rates have been low and corrections have been minimal.[20]

QUALITY MONITORING

Even though every effort is made to obtain high quality data, a monitoring or surveillance system is crucial. When errors are found, a monitoring system enables the investigator to take corrective action. In order to accomplish this, monitoring needs to be current. What is more, monitoring allows an assessment of data quality when interpreting study results. Numerous forms and procedures can be monitored, but doing so is usually not feasible. Rather, monitoring those areas most important to the trial is recommended. Form completion, procedures and drug handling also need to be monitored.

During the study, all forms should be checked for completeness, internal consistency and consistency with other forms. On a follow-up visit to evaluate a subject's progress, the investigator might want to know whether the subject has had a myocardial infarction since the previous follow-up visit. If the subject has had such an event, then more information about the infarction can be collected by completing a special event form. The number of nonfatal myocardial infarctions listed on follow-up visit forms and special event forms should agree. When the forms disagree, the person or group responsible for ensuring consistent and accurate forms should question the person filling out the forms. Consistency within a given form can also be evaluated. Dates and times are particularly prone to error.

It may be important to look at consistency of data over time. A subject with a missing leg on one examination was reported to have palpable pedal pulses on a subsequent examination. Cataracts which did not allow for a valid eye examination at one visit were not present at the next visit, without surgery having been performed. The data forms may indicate extreme changes in body weight from one visit to the next. In such a case, changing the data after the fact is likely to be inappropriate because the correct weights may be unknown. However, the investigator can take corrective action for future visits by more carefully training his staff. Sometimes, mistakes can be corrected. In one trial, comparison of successive electrocardiographic readings disclosed gross discrepancies in the coding of abnormalities. The investigator discovered that one of the technicians responsible for coding the electrocardiograms was fabricating his readings. In this instance, correcting the data was possible.

Someone needs constantly to monitor completed forms to find evidence of missing subject visits, or visits that are off schedule in order to correct any problems. Frequency of missing or late visits may be associated with the intervention. Differences between groups in missed visits may bias the study results. Monitoring and editing of forms is often insufficient and, to improve upon results, it may be necessary to observe actual clinic procedures. Observing clinic procedures is particularly important in multicenter trials.[21]

Extreme laboratory values should also be checked. Values incompatible with life (eg, potassium of 10 mEq/1) are obviously incorrect. Other, less extreme values (ie, cholesterol of 125 mg/dl in male adults in the United States) should be questioned. They may be correct, but it is unlikely. Finally, values should be compared with previous ones from the same subject. Certain levels of variability should be present, but when these levels are exceeded, the value should be flagged. For example, unless the study involves administering a lipid-lowering therapy, any determination which shows a change in serum cholesterol of perhaps 20% or more from one visit to the next should be repeated. Repetition would require saving samples of serum until the analysis has been checked. As well as checking results, a helpful procedure is to monitor submission of laboratory specimens to ensure that missing data are kept to a minimum.

Investigators doing special procedures (laboratory work, electrocardiogram reading) need to have an internal quality control system. Such a system should include re-analysis of duplicate specimens or materials at different times in a blinded fashion. A system of resubmitting specimens from outside the laboratory or reading center might also be instituted. As noted by McCormick and colleagues,[5] these specimens need to be indistinguishable from actual study specimens. An external surveillance system, which should be established in the planning phase of a trial, can pick up errors at many stages (specimen collection, preparation, transportation, and reporting of results), not just at the analysis stage. Thus, it provides an overall estimate of quality. Unfortunately, the system most often cannot indicate at which step in the process errors may have occurred. The external quality control programs implemented in the Coronary Drug Project have been described by Canner et al.[22] Another example is provided by the National Cooperative Gallstone Study.[23]

All recording equipment should be checked periodically. Even though initially calibrated, the machines can break down or require adjustment. Scales can be checked by means of standard weights. If aneroid sphygmomanometers are used, they should be compared regularly with a mercury sphygmomanometer. Factors such as linearity, frequency response, paper speed and time constant should be checked on electrocardiograph machines. In one long-term trial, the prevalence of specific electrocardiographic abnormalities was monitored. The sudden appearance of a three-fold increase in one abnormality, without any obvious medical cause, led the investigator to correctly suspect electrocardiograph machine malfunction.

In a drug study, the quality of the drug preparations should be monitored throughout the trial. Monitoring, as such, includes periodically examining containers for possible mislabeling and for proper contents (both quality and quantity). It has been reported[24] that in one trial, "half of the study group received the wrong medication" due to errors at the pharmacy. Investigators should carefully look for discoloration and breaking or crumbling of capsules or tablets. When the agents are being prepared in several batches, samples from each batch should be examined and analyzed. Occasionally, monitoring the number of pills or capsules per bottle is useful. The actual bottle content of pills should not vary by more than one or two percent. The number of pills in a bottle is important to know because pill count may be used to measure compliance of subjects.

Another aspect to consider is the storage shelf life of the preparations and whether they deteriorate over time. Even if they retain their potency, do changes in odor (as with aspirin) or color occur? If shelf life is long, preparing all agents at one time will minimize variability. Of course, in the event that the study ends prematurely, there may be a large supply of unusable drugs. Products having a short shelf life require frequent production of small batches. Complete records should be maintained for all drugs prepared, examined and used. Ideally, a sample from each batch should be saved. After the study is over, questions about drug identity or purity may arise and samples will be useful.

The dispensing of medication should also be monitored. Checking has two aspects. First, were the proper drugs sent from the pharmacy or pharmaceutical company to the clinic? If the study is double-blind, the clinic staff will be unable to check on this. They must assume that the medication has been properly coded. However, in unblinded studies, staff should check to assure that the proper drugs and dosage strengths have been received. In one case, the wrong strength of potassium chloride was sent to the clinic. The clinic personnel failed to notice the error. An alert subject to whom the drug was issued brought the mistake to the attention of the investigator. Had the subject been less alert, serious consequences could have arisen. An investigator has the obligation to be as careful about dispensing drugs as is a licensed pharmacist. Close reading of labels is essential, as well as documentation of all drugs that are handed out to subjects.

Second, when the study is blinded, the clinic personnel need to be absolutely sure that the code number on the container is the proper one. Labels and drugs should be identical except for the code; therefore, extra care is essential. If bottles of coded medication are lined up on a shelf, it is relatively easy to pick up the wrong bottle accidentally. Unless the subject notices the different code, such errors may not be recognized. Even if he is observant, he may assume that he was meant to receive a different code number. The clinic staff should be asked to note on a study form the code number of the bottle dispensed and the code number of bottles that are returned by the subject. Theoretically, that should enable investigators to spot errors. In the end, however, investigators must rely on the care and diligence of the staff person dispensing the drugs.

It may be worthwhile periodically to send study drug samples to a laboratory for analysis. Although the center responsible for packaging and labeling drugs should have a "foolproof" scheme, independent laboratory analysis serves as an additional check on the labeling process.

The drug manufacturer assigns lot, or batch, numbers to each batch of drugs that are prepared. If contamination or problems in preparation are detected, then only those drugs from the problem batch need to be recalled. This is especially important in clinical trials, since the recall of all drugs can severely delay, or even ruin, the study. When only some drugs are recalled, the study can usually manage to continue. Therefore, the lot number of the drug as well as the name or code number should be listed in the subject's study record.

Finally, monitoring of data quality proves most valuable when there is feedback to the clinic staff and technicians. Once weaknesses and errors have been identified, performance can be improved. Chapter 19 contains several tables illustrating quality control reports. With careful planning, reports can be provided and improvement can be accomplished without unblinding the staff. All quality control measures take time and money; it is thus impossible to be compulsive about the quality of every piece of datum and every procedure. Investigators need to focus their efforts on those procedures which yield key data; those on which the conclusions of the study critically depend.

REFERENCES

1. Koran, L.M. The reliability of clinical methods, data and judgments. Part 1. *N Engl J Med*. 293:642–646, 1975.
2. Koran, L.M. The reliability of clinical methods, data and judgments. Part 2. *N Engl J Med*. 293:695–701, 1975.
3. Belk, W.P., and Sunderman, F.W. A survey of the accuracy of chemical analyses in clinical laboratories. *Am J Clin Pathol*. 17:853–861, 1947.
4. Lewis, S.M., and Burgess, B.J. Quality control in haematology: report of interlaboratory trials in Britain. *Br Med J*. 4:253–256, 1969.
5. McCormick, W., Ingelfinger, J.A., Isakson, G. et al. Errors in measuring drug concentrations. *N Engl J Med*. 299:1118–1121, 1978.
6. Feinstein, A.R., Gelfman, N.A., Yesner, R. et al. Observer variability in the histopathologic diagnosis of lung cancer. *Am Rev Respir Dis*. 101:671–684, 1970.
7. Davies, L.G. Observer variation in reports on electrocardiograms. *Br Heart J*. 20:153–161, 1958.
8. Smyllie, H.C., Blendis, L.M., and Armitage, P. Observer disagreement in physical signs of the respiratory system. *Lancet* 2:412–413, 1965.
9. Assaad, F.A., and Maxwell-Lyons, F. Systematic observer variation in trachoma studies. *Bull WHO* 36:885–900, 1967.
10. Birkelo, C.C., Chamberlain, W.E., Phelps, P.S. et al. Tuberculosis case finding: a comparison of the effectiveness of various roentgenographic and photofluorographic methods. *JAMA* 133:359–366, 1947.
11. Detre, K.M., Wright, E., Murphy, M.L. et al. Observer agreement in evaluating coronary angiograms. *Circulation* 52:979–986, 1975.
12. Frieden, J., Shapiro, J.H. and Feinstein, A.R. Radiologic evaluation of heart size in rheumatic heart disease. *Arch Intern Med*. 111:90–96, 1963.
13. Rose, G.A., Holland, W.W., and Crowley, E.A. A sphygmomanometer for epidemiologists. *Lancet* 1:296–300, 1964.

14. Wilcox, J. Observer factors in the measurement of blood pressure. *Nurs Res.* 10:4–20, 1961.

15. Rose, G.A. and Blackburn, H. *Cardiovascular Survey Methods.* Geneva: World Health Organization, Monograph Series No. 56, 1968.

16. Kahn, H.A., Leibowitz, H., Ganley, J.P. et al. Standardizing diagnostic procedures. *Am J Ophthalmol.* 79:768–775, 1975.

17. Russell, M.L., Ghee, K.L., Probstfield, J.L., et al. Development of standardized simulated patients for quality control of the clinical interview. *Controlled Clin Trials* 4:197–208, 1983.

18. Knatterud, G.L., Forman, S.A., and Canner, P.L. Design of data forms. *Controlled Clin Trials* 4:429–440, 1983.

19. Wright, P., and Haybittle, J. Design of forms for clinical trials. *Br Med J.* 2:529–530, 590–592, 650–651, 1979.

20. Bagniewska, A., Black, D., Curtis, C., et al. Data quality control in a distributed data processing system: nature and method. (Abstract). *Controlled Clin Trials* 4:148, 1983.

21. Ferris, F.L., and Ederer, F. External monitoring in multiclinic trials: applications from ophthalmologic studies. *Clin Pharmacol Ther.* 25:720–723, 1979.

22. Canner, P.L., Krol, W.F., and Forman, S.A. External quality control programs. *Controlled Clin Trials* 4:441–466, 1983.

23. Habig, R.L., Thomas, P., Lippel, K., et al. Central laboratory quality control in the National Cooperative Gallstone Study. *Controlled Clin Trials* 4:101–123, 1983.

24. Quality assurance of clinical data: Discussion. *Clin Pharmacol Ther.* 25:726–727, 1979.

CHAPTER 11

Assessment and Reporting
of Adverse Effects

The assessment of adverse effects encompasses the whole spectrum of research from laboratory work during drug development, animal studies, and early work in small numbers of human beings, to case reports, clinical trials, and post-marketing surveillance. Carcinogenic or teratogenic adverse effects, such as noted with diethylstilbestrol[1,2] and thalidomide,[2,3] receive considerable publicity, but other sorts of findings are undoubtedly more common.

The discussion here is limited to assessment of adverse effects in drugs beyond the initial stages of development and testing. That is, even though the drugs may not yet be marketed, they have undergone early evaluation in human beings and are ready for larger scale evaluation. For the purposes of this book, adverse effects are defined as any clinical event, sign, or symptom that goes in an unwanted direction. It encompasses any undesired outcome of evaluated clinical occurrences (that is, the opposite of primary or secondary response variables), but has the added dimension of physical findings, complaints, and laboratory results. Adverse effects can include both objective measures and subjective responses.

The assessment and reporting of adverse effects have often played, and continue to play, a secondary role in clinical trials. Adverse effects are considered to be important and most reports of studies present data on selected ones. However, trials are generally not designed for the purpose of assessing adverse effects. The scientific standards which are used in evaluating an intervention for efficacy are rarely employed when evaluating possible adverse effects.

Unlike the few and generally well-defined primary or even secondary response variables, many possible adverse effects may be monitored in a trial. Because of this problem of multiple testing (Chapter 15), the intervention may appear significantly different from the control more often by chance alone than indicated by the P-value. Nevertheless, investigators tend to relax their statistical requirements for declaring adverse effects to be real findings. It reflects understandable conservatism and the desire to avoid unnecessary harm.

Adverse effects can be both expected or unexpected. Expected ones are those which, based on previous knowledge about the intervention or similar drugs, are known or likely to occur. Occasionally, completely unanticipated problems can occur. These may be serious enough to lead to termination of the trial, or even to withdrawal of the agent from the market.[4]

In assessing adverse effects, it must be noted that many problems arise with some frequency in the control group. These may be due to the natural history of the disease or condition or to the non-study therapy the subject is receiving. In the Beta-Blocker Heart Attack Trial[5] of propranolol in subjects with recent myocardial infarction, 66% of the group on placebo complained of shortness of breath over the course of two years. At baseline, only 6% had a history of shortness of breath.[6]

As discussed by Bulpitt,[7] the noting of an adverse effect by subjects in an accurate way depends greatly on the setting and manner in which it is elicited, as well as the phrasing of the question. Good questions must be clear and easily understood, have high repeatability, be valid, and must be answered by the subject. These features apply to all areas in clinical trials, but are perhaps most important when the investigation is dealing with subjective complaints.

This chapter will cover a number of determinants of adverse effects; that is, factors which can influence their observed or reported occurrence, frequency, and severity. It will also review several aspects in the reporting or presentation of adverse effects.

FUNDAMENTAL POINT

Adequate attention needs to be paid to the assessment, analysis, and reporting of adverse effects to permit valid assessment of potential risks of interventions.

DETERMINANTS OF ADVERSE EFFECTS

A number of factors play a role in the determination of adverse effects. These can affect the quality of the data, the reported frequency of the effects, and the interpretation of their meaning.

Definitions

The rationale for defining adverse effects is similar to that for defining any response variable: it enables investigators to record something in a consistent manner. Further, it allows someone reviewing a trial to better assess it, and possibly to compare the results with those of other trials of similar interventions.

Because adverse effects are typically viewed as truly secondary or tertiary response variables, they are not often seriously thought about ahead of time. Generally, an investigator will prepare a list of potential adverse effects on a study form, using, perhaps, commonly accepted terms. These usually are not defined, except by the way investigators apply them in their daily practice. Study protocols seldom contain written definitions of adverse effects, unlike primary or other secondary response variables. In multicenter trials, the situation may often be even worse. In those cases, an adverse effect may be simply what each investigator declares it to be. Thus, intrastudy consistency may be poor.

Given the large number of possible adverse effects, it is not feasible to define clearly all of them. Though it is not always easy, important adverse effects which are associated with individual signs or laboratory findings, or a constellation of signs, symptoms, and laboratory results can and should be well-defined. These include ones known to be associated with the intervention and which are clinically significant. Other adverse effects which are purely based on a subject's report of symptoms, may be important, but are more difficult to define. These may include nausea, fatigue, or headache. Nevertheless, the fact that an adverse effect is not well-defined, should be stated in the study protocol. Some adverse effects cannot be defined in that they are not listed in advance, but are spontaneously mentioned by the subjects. Any trial publication should indicate which events were not prespecified.

Ascertainment

The issue of whether one should elicit adverse effects by means of a checklist or rely on the subject to volunteer complaints often arises. Eliciting adverse effects has the advantage of allowing a standard way of obtaining information on a preselected list of symptoms. Thus, both within and between trials, the same series of effects can be ascertained in the same way, with assurance that a "yes" or "no" answer will be present for each. This presupposes, of course, adequate training in the administration of the questions. Volunteered responses to a question such as "Have you had any health problems since your last visit?" have the possible advantage of tending to yield only the more serious episodes, while others are likely to be ignored or forgotten. In addition, only volunteered responses will give information on truly unexpected adverse effects.

In the Aspirin Myocardial Infarction Study,[8] information on several

150

adverse effects was both volunteered by the subjects and elicited. After a general question about adverse effects, the investigators asked about specific complaints. The results for three adverse effects are presented in Table 11-1. Two points might be noted. First, for each adverse effect, eliciting gave a higher percent of subjects with complaints than did asking for volunteered problems. Second, the same aspirin-placebo differences were noted, regardless of the method. Thus, the investigators could arrive at the same conclusions with each technique. In this study, little additional information was gained by the double ascertainment. Perhaps the range between the volunteered numbers and the solicited within the individual study groups provides bounds on the true incidence of the adverse effect.

Downing et al reported on a comparison of elicited versus volunteered adverse effects in a trial of tranquilizers and antidepressants.[9] Thirty-three subjects on active drug volunteered complaints, as opposed to 12 on placebo. This contrasts with 53 elicited complaints from the active drug group and 12 elicited from the placebo group. The authors concluded that eliciting adverse effects preferentially increases the number in the active drug group, rather than the placebo group. This is contrary to the findings in the Aspirin Myocardial Infarction Study.

In the same paper, Downing et al examined the severity of adverse effects obtained by both methods. Of 29 drug-treated subjects who had complaints ascertained by both eliciting and volunteering, 26 were classified as more severe. Of 24 subjects whose complaints were ascertained only by eliciting, half were called more severe. Therefore, the requirement that an adverse effect be volunteered by a subject led to a preponderance of severe ones.

In a clinical trial of antirheumatic drugs, Huskisson and Wojtulewski also compared elicited ("checklist") adverse effects with volunteered ("no checklist").[10] They created a score from the sums of the subject responses, weighted by perception of adverse effect severity. When the adverse effects were elicited, "the difference between the incidence of the significant side effect and the 'background' incidence in the control population was diminished." In addition, the volunteer approach yielded more unlisted, and presumably unexpected adverse effects than did the elicited approach. Avery and colleagues also compared "checklist" versus "nonchecklist" ascertainment of adverse effects in a series of depressed patients.[11] They came to the opposite conclusion concerning severity; namely, there are "significantly greater numbers and greater severity of side effects reported in the checklist group."

Because of these somewhat inconsistent findings, many researchers have continued to use both methods. Recognizing this, the National Institute of Mental Health has developed a Systematic Assessment for Treatment Emergent Events.[12] This instrument provides a standard way

Table 11-1
Percent of Patients Ever Reporting (Volunteered and Solicited) Selected Adverse Effects, by Study Group, in the Aspirin Myocardial Infarction Study

	Hematemesis	Tarry Stools	Bloody Stools
Volunteered			
Aspirin	0.27	1.34*	1.29†
Placebo	0.09	0.67	0.45
Elicited			
Aspirin	0.62	2.81*	4.86*
Placebo	0.27	1.74	2.99

*Aspirin–placebo difference >2 S.E.
†Aspirin-placebo difference >3 S.E.
Aspirin group: $N = 2267$.
Placebo group: $N = 2257$.

of asking both general and specific items for different organ systems. It allows the investigator to enter both symptoms and physical findings in a systematic way. Evaluation of the utility of this instrument in clinical trial settings is underway.

Number of Events

Assessment of an adverse effect is also affected by the number of subjects who suffer the effect. The more subjects with an occurrence of an effect, the more reliable is the estimate of its true frequency. Adverse effects that occur only rarely will only be detected in very large trials. Another problem with low frequency effects (especially unexpected ones) is that investigators are unsure when to ignore them. Disregarding them may be entirely appropriate. On occasion, however, an important adverse effect may be erroneously overlooked because just one or two subjects were affected.

An unavoidable difficulty in documenting drug-related problems is that investigators calculate sample sizes on the basis of primary response variables, not as a result of estimates of adverse effect frequency. Even if assessment of adverse effects is a major objective of the study, a trial's size will not generally be increased in an effort to improve the likelihood of reliably detecting such effects. A possible exception is a study which aims to show no important difference in efficacy between a new intervention and a standard therapy, which serves as the control. If indeed, no clinically major difference between groups for the primary response variable is demonstrated, the ultimate choice of therapy may depend on factors such as low cost, ease of administration, and lack of serious or annoying adverse effects. In this circumstance, an investigator should consider adverse effect assessment, as well as primary outcome assessment, in the sample size estimate.

Length of Follow-up

Obviously, the duration of a trial has a substantial impact on adverse effect assessment. The longer the trial, the more opportunity one has to discover adverse effects, especially those with low frequency. Also, the number of subjects in the intervention group complaining will increase, giving a better estimate of the adverse effect incidence. Of course, eventually, most subjects will report some complaint, such as headache or fatigue. However, this will occur in the control group as well. Therefore, if a trial lasts for several years, and an adverse effect is analyzed simply on the basis of cumulative number of subjects suffering from it, the results may not be very informative.

Duration of follow-up is also important in that exposure time may be critical. Some drugs may not cause certain adverse effects until a person has been taking them for a minimum period. An example is the lupus syndrome with procainamide. Given enough time, a large proportion of subjects will develop this syndrome, but very few will do so if followed for only several weeks.[13] Other sorts of time patterns may be important as well. Many adverse effects occur at low drug doses shortly after initiation of treatment. In such circumstance, it is useful, and indeed prudent, to carefully monitor subjects for the first few hours or days. If no reactions occur, the subject may be presumed to be at a low risk of developing these effects subsequently. In the Beta-Blocker Heart Attack Trial,[5] all subjects were started on a test dose of 20 mg of propranolol (or matching placebo), before being raised to the early maintenance dose of 40 mg three times a day. The purpose was to discover those subjects unable to tolerate the drug while they were still in a closely monitored hospital setting.

Figure 11-1 illustrates the first occurrence of ulcer symptoms and complaints of stomach pain, over time, in the Aspirin Myocardial Infarction Study.[8] Ulcer symptoms rose fairly steadily in both the aspirin and placebo groups, peaking at 36 months. In contrast, complaints of stomach pain were maximal early in the aspirin group, then decreased. Subjects on placebo had a constant, low level of stomach pain complaints. If a researcher tried to compare adverse effects in two studies of aspirin, one lasting weeks and the other months, his findings would be different. To add to the complexity, the aspirin data in the study of longer duration may be confounded by changes in aspirin dosage and concomitant therapy.

An intervention may cause continued discomfort throughout a trial, and that persistence may be an important feature. Yet, unless the discomfort is considerable, such that the intervention is stopped, the subject may eventually stop complaining about it. Unless the investigator is alert to this possibility, the proportion of subjects with symptoms towards the end of a long-term trial may be misleadingly low.

Kinds of Subjects

Different kinds of people react differently to drugs, both in terms of benefit and adverse effects. Age, sex, stage of disease, and many other features of the subject or his condition can and will affect the incidence and severity of specific effects. In addition, they can alter the perception of or actual importance of the adverse effects. In trials of primary prevention, where the subjects may be entirely asymptomatic, even relatively mild adverse effects may not be tolerated. Subjects with diseases of moderate severity, who hope for recovery or remission, may

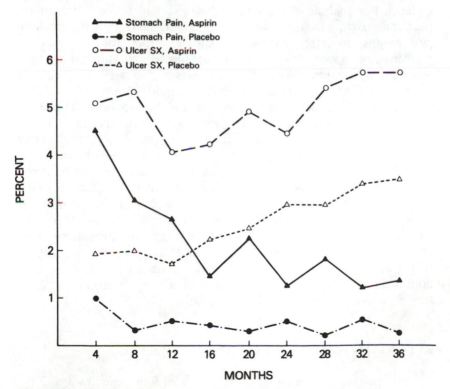

Figure 11-1 Percent of patients reporting selected adverse effects, over time, by study group, in the Aspirin Myocardial Infarction Study. (sx = symptom)

accept more toxic reactions from a drug. While some patients with a terminal illness may accept many bothersome effects and indignities in the hope of a cure, others may refuse anything that increases their discomfort.

The dose and dosing schedule of the study medication can also have different impacts in different subjects. Elderly subjects often require smaller doses than younger subjects. In long-term studies, subjects who are employed outside the home may have difficulties complying with multiple dosing regimens, especially if the drug needs to be taken in the middle of the day. Not only is it inconvenient, but otherwise acceptable adverse effects such as flushing or drowsiness may cause embarrassment if noticed by co-workers.

Subject Withdrawal Policy

As discussed in Chapter 16, there are differing views on analyzing data from subjects who fail to comply with the study intervention regimen. For analysis of primary response variables, the "intention-to-treat" approach, which includes all subjects in their originally randomized groups, is more conservative and less open to bias than the "explanatory" approach, which omits subjects who stop taking their assigned intervention. When adverse effects are assessed, however, the issue is less clear. Subjects are less likely to report adverse effects if they are off medication (active or placebo) than if they are on it. Therefore, the intention-to-treat principle may underestimate the true incidence of adverse effects, by inflating the number of subjects at risk. On the other hand, the explanatory policy makes it impossible to assess events which occur sometime after drug discontinuation, but may in fact be a real adverse effect that was not recognized until later. In addition, withdrawing subjects can void the benefits of randomization, resulting in invalid group comparisons. While there is no easy solution to this dilemma, it is probably safer and more reasonable to continue to assess adverse effects for the duration of the trial, even if a subject has stopped taking his study drug. The analysis and reporting might then be done both including and omitting noncompliant subjects. Certainly, it is extremely important to specify what was done.

REPORTING ADVERSE EFFECTS

Adverse effects can be reported in at least four general ways. First, there may simply be an indication that subjects either did or did not have a particular problem. Second, an investigator may provide an estimate of the severity of an adverse effect. Third, the frequency that the effect occurs in a given subject over a given time period can be presented. Fourth, time patterns may be shown.

The usual measures of adverse effects include:

a) reasons subjects are taken off study medication;
b) reasons subjects are on reduced dosage of study medication;
c) type and frequency of subject complaints;
d) laboratory measurements, including x-rays;
e) in long-term studies, possible drug-related reasons subjects are hospitalized;
f) combinations or variations of any of the above.

All of these can rather easily indicate the number of subjects with a particular adverse effect during the course of the trial. Severity indices can be more complicated. It may be assumed that a subject who was taken off study drug because of an adverse effect had a more serious episode than one who merely had his dosage reduced. Someone who required dose reduction probably had a more serious effect than one who complained, but continued to take the dose required by the study protocol. Data from the Aspirin Myocardial Infarction Study,[8] using the same adverse effects as in the previous example, are shown in Table 11-2. In the aspirin and placebo groups, the percents of subjects complaining about hematemesis, tarry stools, and bloody stools are compared with the percents having their medication dosage reduced for those events. As expected, numbers complaining were many times the numbers with reduced dosage. Thus, the implication is that most of the complaints were for relatively minor occurrences or had been transient.

Table 11-2
Percent of Patients with Drug Dosage Reduced or Complaining of Selected Adverse Effects, by Study Group, in the Aspirin Myocardial Infarction Study

	Aspirin ($N = 2267$)	Placebo ($N = 2257$)
Hematemesis		
Reasons reduced	0	0
Complaints	.27	.09
Tarry Stools		
Reasons reduced	.09	.04
Complaints	1.34	.67
Bloody Stools		
Reasons reduced	.22	.04
Complaints	1.29	.45

Many clinical trials use a severity scale which requires the investigator and/or the subject to indicate on a line or series of boxes the perceived degree of severity.[12,14] It is then possible to indicate the proportion of subjects with a particular adverse effect who have greater than a specified level of severity. The Eastern Cooperative Oncology Group toxicity criteria have five levels of severity for each of about 20 signs or symptoms.[14] Alternatively, a composite score, incorporating several factors, can be created. As previously mentioned, another way of reporting severity is to establish a hierarchy of consequences of adverse effects, such as permanently off-study drug, which is more severe than permanently on reduced dosage, which is more severe than ever on reduced dosage, which is more severe than ever complaining about the effect. Unfortunately, few clinical trial reports present such severity data.

The frequency with which a particular adverse effect has occurred in any subject can be viewed as another measure of severity. For example, episodes of nausea occurring daily, rather than weekly, is obviously more troublesome to the subject. Presenting such data in a clear fashion is complicated. It can be done by means of frequency distributions, but these consume considerable space in tables. Another method is to select a frequency and assume that adverse effects which occur less often in a given time period are less important. Thus, only the number of subjects with a frequency of specified adverse effects above that are reported. As an example, of ten subjects having nausea, three might have it at least twice a week, three at least once a week, but less than twice, and four less than once a week. Only those six having nausea at least once a week might be included in a table. These ways of reporting assume that adequate and complete data have been collected, and may require the use of a diary. Obviously, if a follow-up questionnaire asks only if nausea has occurred since the previous evaluation, frequency measures cannot be presented.

The dimension of time can complicate the reporting of adverse effects. First, subjects can be listed as ever having been off medication, or on reduced doses. Second, they can also be off drug or on reduced dose at some specified time during the trial. This specified time may be the interval between two particular clinic visits. For example, the percent of patients on reduced study medication in the first year of treatment could be presented.

A third time-related way of looking at adverse effects is to indicate the percent of subjects permanently off or taking a reduced dose of study drug. Depending upon the pattern of drug discontinuation and possible re-starting, this measure may or may not provide results similar to the percent of subjects off drug during some specified time period.

As with other response variables, adverse effects can be analyzed using survival analysis (Chapter 14). An advantage of this sort of presentation is that the time to a particular episode, in relation to when the drug was started, is examined. Further, the frequency of a particular adverse effect will be directly related to the number of subjects at risk of suffering it. This can give a higher rate of adverse effects than other measures, but this high rate may be a realistic estimate. Difficulties with survival analysis techniques include the problems of interpreting repeated episodes in any subject and severity of a particular adverse effect, changes in dosing pattern or compliance, and changes in sensitization or tolerance to adverse effects. Nevertheless, this technique has been underutilized in reporting adverse effects.

An attempt to compare adverse effects reported from several studies of similar agents, in this case beta-blockers, has been made.[15] To minimize potential difficulties, only large, placebo-controlled, long-term trials, using approximately equipotent doses of beta-blockers in postmyocardial infarction patients were evaluated. Despite this, different follow-up times, severity of illness of the study subjects, time from acute illness to initiation of study drug, and definitions of adverse effects made review of the published papers difficult. Even after contacting study investigators and obtaining unpublished data, only a few reasons for permanent drug discontinuation could be reasonably compared. No obvious major differences in complications of therapy were discovered, but, given the problems, the conclusions must be interpreted cautiously.

OTHER ISSUES

Clinical trials are excellent mechanisms for comparing groups of subjects and deciding whether a particular intervention is beneficial or harmful. For example, they can be used to show that an intervention causes 25% of the subjects exposed to it to develop nausea, as compared with 10% of subjects on placebo, over the follow-up period. Clinical trials, however, cannot say precisely which particular intervention subjects are among the "expected" 10% and which are among the "excess" 15%. It is even possible that some among the expected 10% with nausea may have had their problem aggravated by the intervention. In general, only individual subject experience, such as drug rechallenge or perhaps documentation of a clear temporal relationship between drug administration and the adverse effect can identify in which subjects the intervention is the responsible agent.

Temple et al[16] and Riegelman[17] indicated other difficulties in using clinical trials to evaluate adverse effects. They concluded that most trials are too small and of too short duration to detect uncommon adverse effects. It has been noted[16] that at most, a few thousand subjects receive a drug during a trial, with perhaps a few hundred receiving it for more than several months. Trials also include a selected type of subject. People most likely to develop adverse effects are generally excluded from clinical trials. Exclusion criteria are intentionally designed to do just that. The thalidomide tragedy[2,3] illustrates the problem that can arise when people who are likely to use a drug are not initially evaluated.

A similar point has been made by Venning,[18] who reviewed the identification and report of 18 adverse effects in a variety of drugs. Clinical trials played a key role in identifying only three of the 18 adverse reactions discussed. Of course, clinical trials may not have been conducted in all of the other instances. Nevertheless, it is clear that assessment of adverse effects, historically, has not been a major contribution of clinical trials.

An important aspect in toxicity assessment is long-term follow-up monitoring of the subjects after the formal trial has ended. Such follow-up is not always feasible in trials which employ continuing drug intervention. They have the possibility for treatment cross-over and contamination of groups. It can be useful, however, if a drug is found to have serious toxicity during the trial. It can then be assumed that few, if any subjects will remain on the drug, or begin taking it. Post-study observation can show whether the adverse effects persist or not. An example of the latter, is the WHO clofibrate trial.[19] There, the post-study attenuation of the observed in-study adverse mortality experience emphasized that the finding was probably real (see Chapter 17 for further discussion). Reasonable assessment of long-term effects of drugs can more easily be accomplished in studies with limited or one-time interventions. In the antenatal steroid therapy trial, the primary response was neonatal respiratory distress syndrome.[20] Even after establishing the short-term impact of therapy, for safety reasons the children were followed for a three-year evaluation of growth and neurologic outcomes.

CONCLUSIONS

Clinical trials have inherent methodological limitations in the evaluation of adverse effects. These include inadequate size, duration of follow-up, and restricted patient selection. Problems in data collection and reporting, however, can be adequately addressed and solved.

The protocol should indicate how adverse effects are defined and collected. Though consistency between studies may not be achieved, what is done should at least be mentioned. Adverse effects should be presented in the main results paper, not buried in subsidiary ones. Published reports should present adverse effect results in several ways. The cumulative number of subjects with a particular problem at any time during the trial and the number with that problem at a particular point are useful overall measures. Some indicator of severity, either using a scale from a questionnaire, or the number of subjects taken off drug, is also important. In long-term trials, the time course of adverse effects should be reported.

Many adverse effects may be assessed. Clearly, not all can be reported. A reasonable approach is to present adverse effects known or suspected to be caused by the intervention, any serious clinical events, and those where a significant difference between groups is observed.

REFERENCES

1. Herbst, A.L., Ulfelder, H., and Poskanzer, D.C. Adenocarcinoma of the vagina. *N Engl J Med*. 284:878–881, 1971.

2. Heinonen, O.P., Slone, D., and Shapiro, S. *Birth Defects and Drugs in Pregnancy*. Littleton: PSG Publishing Company, 1977, pp. 1–7.

3. McBride, W.G. Thalidomide and congenital malformations. *Lancet* ii: 1358, 1961.

4. Multicentre International Study. Improvement in prognosis of myocardial infarction by long-term beta-adrenoreceptor blockade using practolol. *Br Med J*. 3:735–740, 1975.

5. Beta-Blocker Heart Attack Trial Research Group. A randomized trial of propranolol in patients with acute myocardial infarction. 1. Mortality results. *JAMA* 247:1707–1714, 1982.

6. Byington, R.P., for the Beta-Blocker Heart Attack Trial Research Group. Beta-Blocker Heart Attack Trial: design, methods, and baseline results. *Controlled Clin Trials* 5:382–437, 1984.

7. Bulpitt, C.J. *Randomised Controlled Clinical Trials*. The Hague: Martinus Nijhoff, 1983, pp. 214–220.

8. Aspirin Myocardial Infarction Study Research Group. A randomized, controlled trial of aspirin in persons recovered from myocardial infarction. *JAMA* 243:661–669, 1980.

9. Downing, R.W., Rickels, K., and Meyers, F. Side reactions in neurotics: 1. A comparison of two methods of assessment. *J Clin Pharmacology* 10:289–297, 1970.

10. Huskisson, E.C., and Wojtulewski, J.A. Measurement of side effects of drugs. *Br Med J*. 2:698–699, 1974.

11. Avery, C.W., Ibelle, B.P., Allison, B., and Mandell, N. Systematic errors in the evaluation of side effects. *Am J Psychiatry* 123:875–878, 1967.

12. Systematic Assessment for Treatment Emergent Events. National Institutes of Mental Health, Alcohol, Drug Abuse, and Mental Health Administration, Rockville, MD.

13. Woosley, R.L., Drayer, D.E., Reidenberg, M.M., et al. Effect of acetylator phenotype on the rate at which procainamide induces antinuclear antibodies and the lupus syndrome. *N Engl J Med*. 298:1157–1159, 1978.

14. Oken, M.M., Creech, R.H., Tormey, D.C., et al. Toxicity and response criteria of the Eastern Cooperative Oncology Group. *Am J Clin Oncol. (CCT)* 5:649–655, 1982.

15. Friedman, L.F. How do various beta blockers compare in type, frequency and severity of their adverse effects? *Circulation* 67 (suppl I):I-89–I-90, 1983.

16. Temple, R.J., Jones, J.K., and Crout, J.R. Adverse effects of newly marketed drugs (editorial). *N Engl J Med*. 300:1046–1047, 1979.

17. Riegelman, R. Letter to the Editor. *Ann Intern Med*. 100:455, 1984.

18. Venning, G.R. Identification of adverse reactions to new drugs, II. How were 18 important adverse reactions discovered and with what delays? *Br Med J*. 286:289–292, 365–368, 1983.

19. Report of the Committee of Principal Investigators. WHO cooperative trial on primary prevention of ischaemic heart disease with clofibrate to lower serum cholesterol: final mortality follow-up. *Lancet* ii:600–604, 1984.

20. Collaborative Group on Antenatal Steroid Therapy. Effects of antenatal dexamethasone administration in the infant: long-term follow-up. *J Pediatr*. 104:259–267, 1984.

CHAPTER 12

Assessment of Quality of Life

The term "quality of life" is widely used by sociologists, economists, policy makers, and others. However, what is meant by quality of life varies greatly depending on the context. In some settings, it may include such economic components as income, housing, material possessions, environment, working conditions, or availability of public services. The kinds of indices that reflect quality of life from a medical or health viewpoint are very different and would include those that might be influenced not only by diseases, but also by interventions.

Many of the early approaches to assessing quality of life were limited in scope and often disease-specific. Rating scales to classify functional status and determine level of disability were developed for patients with cancer,[1] heart disease,[2] and rheumatoid arthritis.[3] Although these scales were clinically useful, their scientific value was limited. Thus, the categorization of functional status was ambiguous, giving rise to large variability in observers' ratings[4] and the scale increments were not equal, making changes and mean values difficult to interpret. It was not until the 1970s that more comprehensive or so-called "general health status" measures evolved.[5] In addition to evaluation of functional status or "activities of daily living," these also incorporated assessments of mental and social functioning. More recently the trend has again been towards more disease-oriented instruments. One or more components of quality of life were measured in fewer than 5% of cancer trials during the period 1956–1976.[6] The phrase "quality of life" was mentioned in the title of more than 200 reports of cancer treatment published between 1978 and 1980.[7] A review of quality-of-life assessment in cardiovascular clinical trials has recently been published.[8]

For any treatment that does not affect mortality or morbidity, the quality-of-life aspect assumes greater importance. "Life has meaning (depth) as well as length."[9] Schipper[10] emphasizes that the outcome measures of a treatment must not only be relevant for the patient but also perceived to be of functional benefit. A patient may be unwilling to accept treatment despite scientific proof of efficacy, if he sees nothing in it for himself. People attach a different relative importance to each of the components that comprise quality of life, and they vary in acceptance of their discomfort in exchange for possible benefit.

Difficulties in understanding the concept of quality of life,[11] as well as a lack of agreement on definitions and inadequate development of standardized measuring techniques, have hindered progress. Nevertheless, the importance of evaluating various aspects of life quality in medicine is now realized[4,5,7,8,12] and many of the methodologic problems are being solved. Health professionals are increasingly aware of the patient's right to participate actively in deciding treatment choices.

FUNDAMENTAL POINT

Assessment of the effect of intervention on a subject's quality of life may be an important response variable in certain types of clinical trials, particularly those of chronic conditions.

Definition

We have adopted the definition of quality of life proposed by Levine and Croog[11] with only minor modifications. There are two major components: functioning and perceptions (Table 12-1). Other definitions have also been proposed.[7,12]

Functioning The different functions performed by an individual can each be affected by a disease, its treatment, or by numerous other factors. *Social* functioning is crucial and has been considered the major component of quality of life. This includes the ability to interact in a community with family and friends, to work, and to carry our roles as spouse, parent, or sibling. These abilities provide satisfaction, recognition, gratification, and a sense of worth. Studies have shown that isolated persons with a poor social network do less well medically.[13,14]

Table 12-1
Dimensions of Quality of Life

I. *Functioning*
   1. Social (ability to work and interact with family, friends, and community)
   2. Physical (mobility, energy, independence, freedom from symptoms)
   3. Emotional (stability, self-control, freedom from impairments)
   4. Intellectual (ability to think clearly and make decisions including alertness, memory, and confidence)

II. *Perceptions*
   1. Life satisfaction (well-being)
   2. Health status (often compared to others of own age)

Physical functioning or activities of daily living relate to a person's ability to perform daily life activities, both occupational and recreational, eg, mobility (getting up in the morning, moving around, and handling chores in and around the home), independence, and energy. Physical functioning is influenced by symptoms such as weakness, sleeplessness, and dizziness.

The third component is the *emotional* state of a subject. A feeling of stability and self-control is an integral part of this type of functioning. Impairments are reflected in symptoms such as hypochondria, helplessness, irritability, hostility, and guilt. Others, such as anxiety, worry, fear, and depression are often associated with being ill.

Lastly, of considerable value to an individual is his *intellectual* functioning or his ability to think clearly and make decisions for himself. Indicators are, for example, memory, alertness, ability to communicate, and confidence.

An important aspect of the four types of functioning is that they can be partially validated by someone in a person's immediate environment, eg, the spouse.

Perceptions In contrast to functioning, perceptions are totally subjective. For instance, one perception relates to a subject's *sense of general satisfaction*. It serves as an integrating or general measure of the functions discussed above, the symptoms associated with the illness and its treatment, and other factors. Another considers the subject's *perception of health*. Perceived and actual health are sometimes confused. A person who is ill and perceives himself as such may, after a period of acceptance, reset his expectations within his adjusted limits and enjoy life. In contrast, a totally healthy person may be dissatisfied with his situation, resulting in poor quality of life. It is interesting to note that perceived health ratings are strongly and independently associated with increased risk of mortality. The relative risk for all-cause mortality for adults who, by response to a single question ("Compared to others your own age, how do you rate your health?"), perceived their health as poor, as compared to individuals who rated themselves as excellent, has been reported to be about 2 or 3, even after adjusting for other factors, including age and objective health status.[15,16] Such self-ratings of health are influenced by age, sex, socioeconomic and employment status, history of chronic diseases, emotional problems, and high levels of stress and life change.[17]

Adverse effects of drugs also influence a person's quality of life. In a clinical trial it is common at each clinic visit to note spontaneously volunteered or directly elicited adverse experiences. As pointed out in Chapter 11, quantifying adverse effects as well as predicting their impact on individual subjects is difficult. Evaluating specific drug-related adverse effects and a person's ability to function, including his perception of health and life satisfaction, can be complementary.

Factors That May Influence Quality of Life

1. Any study intervention may induce changes, improvements as well as impairments, in a subject's quality of life. One objective of a trial may be to determine which changes are induced by intervention and which are the result of other factors discussed below.

2. The underlying disease can also affect a subject's quality of life. It has been claimed that the mere labeling of somebody as hypertensive may lead to disability and mood changes.[18,19] The knowledge of having a serious illness is sufficient to impair quality of life for most people. In addition, the reduced capacity to work because of severe disease may result in unforeseen financial hardship with all its adverse consequences.[7] Collection of data at baseline is, therefore, important. This information characterizes the subjects' status at entry to the study, before the intervention program is started; it may also be used to assess group comparability. If certain quality-of-life factors have prognostic significance, stratified randomization may be considered.

3. Changes in the natural course of the disease or condition must be considered especially in trials of longer duration. These changes that are unrelated to intervention are likely to occur in all study groups. If prevalent, they may make it difficult to ascertain intervention-induced changes.

4. Concomitant interventions or the regimen of care itself may also affect quality of life. This is particularly likely to happen in trials in which the visit schedule for subjects assigned to active intervention is different from that for the control group. Trials with an open, or unblinded, design are susceptible to the same problem. Subjects in one group may get preferential compensatory treatment (Chapter 6) or more "tender loving care."

5. Life events unrelated to the illness or intervention, eg, bereavement or a sweepstakes' win, could also change a person's outlook.

The majority of publications on quality-of-life measures are limited in scientific value because the studies were uncontrolled. Subjects have been assessed either at one point in time or before and after intervention and it is impossible to be certain to what extent a deviation or change in the components of quality of life is intervention-related or due to other factors. A randomized, control group is necessary if the specific effects of an intervention on quality of life are to be adequately assessed.

Uses of Quality-of-Life Assessment in Clinical Trials

There are two general reasons for assessing quality of life in a clinical trial. First, aspects of quality of life might, in certain situations, be important primary or secondary response variables. Although, the purpose of therapy may be to ameliorate an existing condition, the result of treatment may be improved, unchanged, or impaired quality of life. Adverse effects of drugs and complications of surgery can have negative consequences. To complicate matters further, it is possible that in any individual, certain components of a quality-of-life assessment may be influenced favorably while others are affected unfavorably. Second, measures of life quality may be predictive of subsequent mortality and morbidity. Ruberman et al[14] recently reported that survivors of acute myocardial infarction classified as being socially isolated and unable to cope with high life stress had a more than fourfold risk of death over a 25-month period compared to others with low levels of both isolation and stress. Measurement of such predictive factors at baseline may be relevant to the assessment of comparability between the study groups (Chapter 8). If these associations are causal, they may be amenable to behavioral intervention. In addition, the experience from a limited number of trials in the cardiovascular and cancer fields[20,21] suggests that assessment of quality of life may in itself have some therapeutic value.

There are four types of trials where indices of quality of life may assume an important role in the overall assessment of an intervention. First, quality-of-life evaluation may be critical in trials where the intervention has a potential effect on symptoms without any expected effect on mortality or rate of complications. Improved functioning and/or perception of improvement could serve as indicators of benefit. This is an important consideration in those medical specialities where the major goal of treatment is to preserve and restore physical ability, eg, in rheumatology.[22] Mood changes with secondary alterations of many functional capabilities are central in psychiatry. Many of the methods available to determine and quantify mood changes have been developed in psychiatric populations.[23]

Second, assessment of quality of life is of special importance when the interventions cause a high frequency of extremely unpleasant, adverse reactions, such as cancer therapy. It has been proposed by investigators at the National Cancer Institute that assessment of health status or quality of life be included in all clinical trials conducted by the National Institutes of Health.[24]

Third, the assessment of these factors is of importance in many prevention trials, particularly when the untreated subjects are relatively asymptomatic, mortality and morbidity rates are low, and the imposed intervention may have undesirable effects. Examples are mild hypertension and hyperlipidemia.

Fourth, quality-of-life measures may be important in trials of new drugs that have little to offer in terms of benefit when compared to standard therapy but may have other advantages such as fewer adverse effects or lower costs.

Quality-of-Life Indices as Outcome Variables

There are a number of examples of randomized control trials in which a quality-of-life assessment was used to evaluate outcome. Kane et al[25] examined several outcomes of hospice care in 247 terminally ill cancer patients. The control subjects received conventional care. The authors reported no statistical differences in measures of pain, other symptoms, affect, or activities of daily living. However, the hospice patients expressed more satisfaction with the care they received and the people caring for them showed less anxiety.

In the Danish Obesity Project,[26] 202 individuals with gross obesity were randomized to either surgical treatment (end-to-side jejunoileostomy) or medical management. Three percent of the 130 subjects receiving surgery had pulmonary complications, 6% had wound infection or dehiscence, and 2% had transient hepatic dysfunction. The median weight loss at 24 months was 42.9 kg in the surgical group compared to 5.9 kg in the medical group. As massive obesity represents a severe handicap, quality of life was assessed by questionnaire. Surgery led to improved social, physiological, and emotional functioning; in particular, it led to increased self-esteem, improved social contacts, and increased work capacity.

Sugarbaker et al[27] tested the hypothesis that limb-sparing surgery plus irradiation would profoundly improve quality of life compared to amputation in 26 patients with soft-tissue sarcoma. The data covering a range of measures of functions (social, daily life, and psychological) failed to reveal substantial group differences. In fact, there were suggestions that the limb-sparing approach was inferior to amputation.

The effect of a comprehensive rehabilitation program in ambulatory persons with rheumatoid arthritis was compared to standard care in a small randomized trial.[28] Changes in physical, psychological, and social functioning were determined as well as measures of disease activity. Although not all of these differences reached statistical significance, the subjects in the intervention group appeared to be better off after one year than the controls, with respect to activities of daily living, socioeconomic functioning, house-confinement, and disease activity.

Coronary artery bypass graft surgery was compared to medical therapy in 780 subjects with stable coronary heart disease.[29] Compared with those receiving medical treatment, patients undergoing surgery experienced less exercise-induced angina and were able to exercise longer, but no difference in survival was found. There was also no difference with respect to employment or recreational status after five years of follow-up.

In a trial of 258 postinfarction subjects, Ott et al[30] tested the benefits of an exercise program alone, a combined exercise and teaching-counseling program, and conventional care (the control group). Rehabilitation alone resulted in limited physical benefit. The subjects in the combined exercise-teaching program, however, showed significant improvement in the psychosocial dimension, in contrast to those in the other two groups. This included increased social interaction, alertness, and improved emotional status.

Methodologic Issues

General The rationale for a well-designed and conducted randomized clinical trial to assess quality-of-life measures is the same as for other response variables. Because the data are subjective, special precautions are necessary. A control group allows the investigator to determine which changes can be reasonably attributed to the study intervention. The double-blind design minimizes the effect of investigator bias. The findings will be all the more credible if hypotheses are established *a priori* in the trial protocol.

The basic principles of data collection (Chapter 10) which ensure that the data are of the highest quality are also applicable. The criteria to be used and the methodology for assessment must be clearly defined. Training sessions are advisable since many investigators may doubt the value of collecting behavioral data. Pretesting of forms and questionnaires may enhance user and patient acceptability. An ongoing monitoring or surveillance system enables prompt corrective action when errors and other problems are found.

Specific Specific issues include several pragmatic considerations such as the time of administration of the questionnaire or interview, the need for any special equipment or training, and the reliability and stability of the scale.[5,7,31-33] Acceptability is essential as poor completion rates can invalidate the study. However, in general, volunteers for a clinical trial are cooperative, so this problem may be smaller than reported in population surveys. The purpose of the procedure has to be explained, sensitive items avoided, and the time requirement kept reasonable. In large-scale trials, self-administered questionnaires may be the only feasible data-gathering method, although in smaller trials, interviews or direct observation may be considered. The difference in cost between a self-administered questionnaire and a personal interview is considerable. More severely ill and elderly persons, as well as those with low educational levels, require more time and assistance.

Many investigators have made the mistake of adopting a questionnaire developed for another population only to find that the distribution of responses obtained is skewed. In part, this may be due to volunteers for a trial comprising a select group, often healthier than patients in general with the same diagnosis. This emphasizes the need to pretest any proposed instrument before a trial.

The control group is the appropriate reference for any comparison to be made. Changes in values in the enrolled subjects rather than absolute scores are used in any group comparison. Different components of a quality-of-life assessment may be combined into a summary score, or alternatively displayed as a health status profile. The first approach, although simple, has disadvantages. It is difficult to know what weight the different components should be given. Moreover, a moderate change in one index induced by an intervention might be lost when all the components are aggregated in a composite score.

Two different concepts have been utilized to assess functioning.[34] One considers the capacity to perform a particular task or activity. The second addresses whether the function is actually performed. From a quality-of-life perspective, what matters is the importance the person attaches to that function. For a physically active individual, being unable to continue a certain recreational activity may be a major loss. For someone physically inactive or without that interest, the inability may be of minor concern.

Usually, since little is known about the impact of any intervention on the different components of a quality-of-life assessment, it may be wise to include in the selection of instruments both specific and broad-based ones. It would be of value for this field if core instruments were generally accepted and applied in clinical trials. This would allow comparisons between similar trials and across studies in different fields.

Sources of Information

In an observational study, Jachuck et al[35] evaluated the quality of life in 75 people with hypertension controlled by drugs. Their physicians judged that all the patients had improved. These opinions were based on the fact that the blood pressure was controlled, that no clinical deterioration had been observed, and that the patients had not complained to the doctors about any ill effects of the drug treatment. However, relatives reported that three quarters of the patients had suffered moderate or severe impairment of their quality of life after receiving the antihypertensive therapy. The ratings from the relatives showed changes in all four areas of functioning—social, physical, emotional, and intellectual. The patients rated their own quality of life, assigning values intermediate between ratings done by their physicians and their relatives. It has been suggested that patients underreport problems in an attempt to please their physicians.

This example shows that the source of information used in assessing the effect of an intervention on a person's quality of life is relevant. It suggests that for these kinds of data physicians may be less than ideal. Although the rating given by the relatives ought to be considered, the most important source is the patient. It is of interest to note that two commonly used measures, the Index of Well-Being and the Sickness Impact Profile, used a panel of judges to generate weight for each item of the instruments.[12]

Interpretation of Quality-of-Life Data

In any evaluation of the effect of an intervention on an individual, the weighing of benefit against harm is crucial. The problem is illustrated by the following quote: "Even if we had the data, however, it would take the wisdom of Solomon to decide whether some price like an average of 20 man-years of impotence for each of 100 men who were young when therapy began was too big a price to pay for preventing two fatal myocardial infarctions in middle age."[36]

Bush and colleagues[37,38] introduced the concept of health-related quality of life or "well-years." The Well-Life Expectancy is the product of the Quality of Well-Being score and the expected duration of disability over a standard life period. This approach allows comparisons of groups of individuals with different severity of the same condition as well as with different illnesses. However, there are several shortcomings with the method.[12]

In certain circumstances, the patients may be closely involved with weighing harm and benefit. McNeil et al[39] analyzed the tradeoffs between quality and quantity of life in laryngeal cancer. Using the principles of expected utility theory to develop a method for sharpening decisions, they investigated 37 healthy volunteers. The results suggest that most were willing to give up survival time in order to avoid losing normal speech if the survival time was above five years. The authors concluded that survival was not the only consideration. It is important to take into account persons' attitudes toward morbidity and to realize that opinions differ greatly from individual to individual.

REFERENCES

1. Karnofsky, D.A., and Burchenal, J.H. The clinical evaluation of chemotherapeutic agents in cancer. Edited by C.N. MacLeod. In *Agents in Cancer*. New York: Columbia University Press, 1949.

2. New York Heart Association: *Diseases of the Heart and Blood Vessels: Nomenclature and Criteria for Diagnosis*. Boston: Little, Brown, 1964.

3. Steinbrocker, O., Traeger, C.H., and Batterman, R.C. Therapeutic criteria in rheumatoid arthritis. *JAMA* 140:659–662, 1949.

170

4. Hutchinson, T.A., Boyd, N.F., and Feinstein, A.R. Scientific problems in clinical scales, as demonstrated in the Karnofsky index of performance status. *J Chronic Dis*. 32:661–666, 1979.

5. Deyo, R.A. Measuring functional outcomes in therapeutic trials for chronic disease. *Controlled Clin Trials* 5:223–240, 1984.

6. Bardelli, D., and Saracci, R. Measuring the quality of life in cancer clinical trials: a sample survey of published trials. *UICC (International Union Against Cancer) Technical Report Series* 36:75–94, 1978.

7. Fayers, P.M., and Jones, D.R. Measuring and analysing quality of life in cancer clinical trials: A review. *Stat Med*. 2:429–446, 1983.

8. Wenger, N.K., Mattson, M.E., Furberg, C.D., and Elinson, J. (Eds): *Assessment of Quality of Life in Clinical Trials of Cardiovascular Therapies*. New York: LeJacq Publishing Inc., 1984.

9. Eiseman, B. The second dimension. *Surgery* 116:11–13, 1981.

10. Schipper, H. Why measure quality of life? *Can Med Assoc J*. 128:1367–1370, 1983.

11. Levine, S., and Croog, S.H. What constitutes quality of life? A conceptualization of the dimensions of life quality in healthy populations and patients with cardiovascular disease. Edited by N.K. Wenger, M.E. Mattson, C.D. Furberg, and J. Elinson. In *Assessment of Quality of Life in Clinical Trials of Cardiovascular Therapies*. New York: LeJacq Publishing Inc., 1984.

12. Jette, A.M. Health status indicators: Their utility in chronic-disease evaluation research. *J Chronic Dis*. 33:567–579, 1980.

13. Berkman, L.F., and Syme, S.L. Social networks, host resistance and mortality: A nine-year follow-up study of Alameda County residents. *Am J Epidemiol*. 109:186–204, 1979.

14. Ruberman, W., Weinblatt, E., Goldberg, J.D., and Chaudhary, B.S. Psychosocial influences on mortality after myocardial infarction. *N Engl J Med*. 311:552–559, 1984.

15. Mossey, J.M., and Shapiro, E. Self-rated health: A predictor of mortality among the elderly. *Am J Pub Health* 72:800–808, 1982.

16. Kaplan, G.A., and Camacho, T. Perceived health and mortality: A nine-year follow-up of the human population laboratory cohort. *Am J Epidemiol*. 117:292–304, 1983.

17. Garrity, T.F., Somes, G.W., and Marx, M.B. Factors influencing self-assessment of health. *Soc Sci Med*. 12:77–81, 1978.

18. Haynes, R.B., Sackett, D.L., Taylor, D.W., et al: Increased absenteeism from work after detection and labelling of hypertensive patients. *N Engl J Med*. 299:741–744, 1978.

19. Report of the Working Group: Mild Hypertension. Edited by N.K. Wenger, M.E. Mattson, C.D. Furberg, and J. Elinson. In *Assessment of Quality of Life in Clinical Trials of Cardiovascular Therapies*. New York: LeJacq Publishing Inc., 1984.

20. Mann, A.H. Factors affecting psychological state during one year on a hypertension trial. *Clin Invest Med*. 4:197–200, 1981.

21. van Dam, F.S.A.M., Linsen, C.A.G., and Couzijn, A.L. Evaluating "quality of life" in cancer clinical trials. Edited by M.E. Buyse, M.J. Staquet, and R.J. Sylwester. In *Cancer Clinical Trials: Method and Practice*. Oxford: Oxford University Press, 1984.

22. Liang, M.H., Cullen, K.E., and Larson, M.G. Measuring function and health status in rheumatic disease clinical trials. *Clin Rheum Dis*. 9:531–539, 1983.

23. Burdock, E.T. (Ed). *The Behavior of Psychiatric Patients: Quantitative*

Techniques for Evaluation. New York: Marcel Dekker Inc., 1982.

24. Barofsky, I., and Sugarbaker, P.H. Health status indexes: disease specific vs. general population measures. In *The Public Health Conference on Records and Statistics. The People's Health: Facts, Figures, and the Future*. U.S. Dept. of Health, Education, and Welfare Publication No. (PHS) 70-1214, pp. 263-269, 1979.

25. Kane, R.L., Wales, J., Bernstein, L., et al. A randomised controlled trial of hospice care. *Lancet* i:890-894, 1984.

26. The Danish Obesity Project. Randomised trial of jejunoileal bypass versus medical treatment in morbid obesity. *Lancet* ii:1255-1258, 1979.

27. Sugarbaker, P.H., Barofsky, I., Rosenberg, S.A., and Gionola, F.J. Quality of life assessment of patients in extremity sarcoma clinical trials. *Surgery* 91:17-23, 1982.

28. Katz, S., Vignos, P.J., Jr., Moskowitz, R.W., et al. Comprehensive outpatient care in rheumatoid arthritis. A controlled study. *JAMA* 206:1249-1254, 1968.

29. CASS Principal Investigators and Their Associates. Coronary Artery Surgery Study (CASS): A randomized trial of coronary artery bypass surgery. Quality of life in patients randomly assigned to treatment groups. *Circulation* 68:951-960, 1983.

30. Ott, C.R., Sivarajan, E.S., Newton, K.M., et al. A controlled randomized study of early cardiac rehabilitation: The Sickness Impact Profile as an assessment tool. *Heart and Lung* 12:162-170, 1983.

31. Liang, M.H., Cullen, K., and Larson, M. In search of a more perfect mousetrap (health status or quality of life instrument). *J Rheum.* 9:775-779, 1982.

32. Ware, J.E., Jr., Brook, R.H., Davies, A.R., and Lohr, K.N. Choosing measures of health status for individuals in general populations. *Am J Pub Health* 71:620-625, 1981.

33. Ware, J.E., Jr., Methodological considerations in the selection of health status assessment procedures. Edited by N.K. Wenger, M.E. Mattson, C.D. Furberg, and J. Elinson. In *Assessment of Quality of Life in Clinical Trials of Cardiovascular Therapies*. New York: LeJacq Publishing Inc., 1984.

34. Bush, J.W. Relative preferences versus relative frequencies in health-related quality of life evaluation. Edited by N.K. Wenger, M.E. Mattson, C.D. Furberg, and J. Elinson. In *Assessment of Quality of Life in Clinical Trials of Cardiovascular Therapies*. New York: LeJacq Publishing Inc., 1984.

35. Jachuck, S.J., Brierly, H., Jachuck, S., and Willcox, P.M. The effect of hypotensive drugs on the quality of life. *J R Coll Gen Pract.* 32:103-105, 1982.

36. Perry, M. in Mild hypertensives in the Hypertension Detection and Follow-up Program and Evaluation of Drug Treatment in Mild Hypertension: VA-NHLBI feasibility trial. General discussion. *Annals NY Acad Sci.* 304: 289-292, 1978.

37. Bush, J.W., Chen, M., and Zaremba, J. Estimating health program outcomes using a Markov equilibrium analysis of disease development *Am J Pub Health* 61:2362-2375, 1971.

38. Kaplan, R.M., and Bush, J.W. Health-related quality of life measurement for evaluation research and policy analysis. *Health Psychol.* 1:61-80, 1982.

39. McNeil, B.J., Weichselbaum, R., and Pauker, S.G. Speech and survival. Tradeoffs between quality and quantity of life in laryngeal cancer. *N Engl J Med.* 305:982-987, 1981.

CHAPTER 13

Subject Compliance

The terms compliance and adherence are often used inter-changeably. This book uses the former term. Compliance means following both the intervention regimen and trial procedures (for example, clinic visits, laboratory procedures and filling out forms). Both aspects involve the investigator and his staff as well as the subject. A non-complier is a subject who fails to meet the standards of compliance as established by the investigator. In a drug trial, he may be a subject who takes less than a predetermined amount (eg, 80%) of the protocol dose.

The optimal study from a compliance point of view is one in which the investigator has total control over the subject, the administration of the intervention regimen (a drug, diet, exercise, or other intervention) and follow-up. That situation exists only in animal experiments. Any clinical trial (which, according to the definition in this text, must involve human beings) is likely to have less than 100% compliance with the intervention. There are several reasons for noncompliance. People tend to forget, they develop side effects, and they change their minds regarding participation. Therefore, even studies of a one-time intervention such as surgery or a single medication dose can have noncompliance. In fact, surgical procedures have been reversed. In addition, the subject's disease may deteriorate, thus requiring termination of the study treatment or a switch from control to intervention. For example, in a clinical trial of medical versus surgical intervention, a subject's status may change to the extent that it might be in the best interest of that subject, if assigned to the medical group, to undergo surgery. In the Coronary Artery Surgery Study, an average of 4.7% of subjects assigned to medical intervention had bypass surgery every year.[1]

Obviously, the results of a trial can be affected by noncompliance with the intervention, which leads to underreporting of possible therapeutic as well as toxic effects. Underreporting can undermine even a properly designed study. Monitoring compliance is critical in a clinical trial, as the interpretation of study results must take into account whether the intervention regimen was followed. This chapter will discuss what can be done before enrollment to reduce future compliance problems, how to maintain good compliance during a study, and how to monitor compliance. Readers interested in a more detailed discussion of various issues are referred to the book *Compliance in Health Care* edited by Haynes, Taylor and Sackett.[2] Eraker and colleagues[3] have recently reviewed the field. Most of the information in the compliance area is obtained from the clinical therapeutic situation, and not from the clinical trial setting. Although the difference between patients and volunteer subjects may be important, tending to minimize noncompliance in trials, it is assumed that the basic principles apply to both.

FUNDAMENTAL POINT

Many potential compliance problems can be prevented or minimized before subject enrollment. Once a subject is enrolled, taking measures to enhance and monitor subject compliance is very important.

Since noncompliance with the intervention can have an impact on the power of a trial, realistic estimates of crossovers, dropins and dropouts must be used in calculating the sample size. By a *crossover* is meant a subject who, although assigned to the control group, follows the intervention regimen; or a subject who, assigned to an intervention group, follows either the control regimen or the regimen of another intervention group when more than one intervention is being evaluated. A *dropin* is a special kind of crossover. In particular, the dropin is unidirectional, referring to a person who was assigned to the control group but begins following the intervention regimen. By a *dropout* is meant a person assigned to an intervention group who fails to comply with the intervention regimen. If the control group is either on placebo or on no standard intervention or therapy, the dropout is the equivalent to a crossover. However, if the control group is assigned to an alternative therapy, then a dropout from an intervention group does not automatically begin following the control regimen. Rather, he takes on some characteristics of a person outside the trial. Moreover, in this circumstance, there may be a dropout from the control group. See Chapter 7 for further discussion of the sample size implications of noncompliance.

CONSIDERATIONS BEFORE SUBJECT ENROLLMENT

Before enrollment, several steps can be taken to minimize compliance problems. *Study design* is a key aspect. The shorter the study, the more likely subjects are to comply with the intervention regimen.[4] A study started and completed in one day or during a hospital stay has great advantages over longer trials. Studies in which the subjects are under supervision, such as hospital-based trials, tend to have fewer problems of non-compliance. However, there is a difference between special hospital wards and clinics with trained attendants who are familiar with research requirements, and general wards and clinics, where research experience might not be common or protocol requirements might not be appreciated. Regular hospital staff have many other duties which compete for their attention and they perhaps have little understanding of the need for precisely following a study protocol. In the latter case, explaining the various aspects of a trial to everyone who may become involved in the study is especially important.

Whenever the study uses subjects who will be living at home, the chances for noncompliance increase. Studies of interventions that require changing a habit are particularly susceptible to this hazard.[5] In dietary studies, when the subject's source of food comes only from the hospital kitchen, he is more likely to comply with that regimen than if he bought and cooked his own food. Using a special commissary to supply the food is a possible approach to dietary intervention for people who are living at home. It also allows for blinded design.[6]

Simplicity of intervention is important. Single dose drug regimens are preferable to multiple dose regimens.[7] Many people forget to take every dose at exactly the right time when on a three- or four-times-a-day regimen. Adding multiple types of drugs, maybe with different dosage schedules, adds to the confusion and results in mistakes. Complying with multiple approaches simultaneously poses special difficulties. For example, quitting smoking, losing weight and reducing the intake of saturated fat at the same time require dedicated subjects. Unlike non-obligatory interventions such as drugs, diet, or exercise, surgery, and vaccination have the design advantage, with few exceptions, of enforcing compliance with the intervention.

An important factor in preventing compliance problems before enrollment is *selection of appropriate subjects*. Ideally, only those people likely to follow the study protocol should be enrolled. Numerous studies[2] have been conducted in an effort to determine what factors distinguish good from poor compliers. Socio-demographic, disease and therapy-related, and investigator-related factors are among the many variables that have been studied. Only a limited number have generally been identified as important. One is the subject's belief in his susceptibility to the consequences of the condition or disease being studied, as well as his general feelings of vulnerability.[8] Related to feelings of susceptibility and vulnerability is the subject's perception of the possible repercussions of an illness. If the consequences are serious, compliance with therapy is usually high. An additional factor is the benefit that the subject thinks is probable from the intervention. Higher level of education has been reported to correlate with good compliance.[9] Clinic site in a multicenter trial has also been noted to influence compliance.[10] The importance of age is unclear.[9,10]

It is usually advisable to exclude certain types of people from participation in a trial. These would include persons addicted to drugs or alcohol (unless the treatment of drug or alcohol addiction is being studied), persons who live too far away, or those who are likely to move before the scheduled termination of the trial. Traveling long distances can be an undue burden on disabled people. Those with concomitant disease sometimes have conflicting interests that result in lack of compliance because they have other medicines to take or are participating in other trials. Furthermore, there is the potential for contamination of the study results by these other medicines or trials.

Although convincing evidence is lacking,[11,12] it seems reasonable that a truly *informed subject* is likely to be a better complier. Therefore, for scientific as well as ethical concerns, the subject (or, in special circumstances, his guardian) in any trial should be clearly instructed about the study and told what is expected from him. Sufficient time should be spent with a candidate. He should be encouraged to consult with his family or private physician. A brochure with information concerning the study is often helpful. As an example, the pamphlet used in the Aspirin Myocardial Infarction Study is shown in Figure 13-1.

Where feasible, a run-in period before actual randomization can be used to identify subjects likely to become poor compliers. During the run-in, subjects may be given either active medication or placebo over several weeks or months. This approach was employed in a trial of aspirin and beta-carotene in US physicians.[13] Alternatively, the investigator may instruct prospective subjects to refrain from taking the active agent and then evaluate how well his request was followed. This was done in the Aspirin Myocardial Infarction Study where urinary salicylates were monitored before enrollment. Very few subjects were excluded because of a positive urine test.

MAINTAINING GOOD SUBJECT COMPLIANCE

After enrollment, a number of steps can be taken to maintain good compliance. These are especially critical in studies of long duration. Experienced investigators stay in *close contact*[11] with the subjects early after randomization to get subjects involved and, later, to keep them interested when their initial enthusiasm may have worn off. Between scheduled visits, clinic staff may make frequent use of the telephone and the mail. Sending cards on special occasions such as birthdays and holidays is a helpful gesture. Visiting the subject if he is hospitalized shows concern. The investigator may make notes of what the subject tells

What is the Aspirin Myocardial Infarction Study?

A nationwide collaborative study among 30 centers, sponsored and supported by the National Heart and Lung Institute, NIH, designed to determine whether aspirin will decrease the risk of recurrent heart attacks in patients who have had a previous heart attack. Forty-two hundred men and women across the country will be participating in this massive research program.

Who Can Participate?

Men and women between the ages of 30 and 69 who have had a documented heart attack within the last five years, but who do not have any other major disease.

Your private physician, if you have one, must give his/her approval before you can be admitted to the study, and you will continue under his/her general medical care.

What Happens in the Study?

Your eligibility will be evaluated at two Initial Visits by means of a brief history, physical examination and some laboratory tests, including a chest x-ray and electrocardiogram. Other information from our hospital or clinic records will be looked at to see if they confirm a heart attack. You will also be briefed regarding details of the study and you will then have an opportunity to ask any questions you might have about the study.

If you are eligible and agree to participate in the study, you will be assigned randomly to take either aspirin or a placebo. You will be asked not to use other aspirin or aspirin-containing drugs. You will be given other tablets to use for headaches or fever instead of aspirin.

What About My Time?

After the initial evaluation period, you will be asked to come in to the clinic about every four months at which time you will see a physician and have laboratory analyses done, including an annual ECG and chest x-ray.

How Long is the Study?

The study participants will be followed for three to four years unless results are obtained earlier.

How Does the Medicine Work?

Aspirin is one the most commonly used drugs in the world. The present study is a test of theory that aspirin, perhaps through its effect on blood clotting, may prevent recurrent heart attacks or death from coronary heart disease. Aspirin has, in the dose that will be used, slight but definite side effects. Patients who are known to react adversely to aspirin or who are considered to be at risk of developing gastric ulcers during the course of the study, will not be allowed to enter it. The enrolled patients will be carefully monitored throughout the study, to prevent side effects.

What are the Benefits?

1. Regular medical and laboratory examinations.
2. The medical evaluations will provide your private physician with important information about your health.
3. All services related to the aspirin study are provided at no cost to you.
4. An opportunity for you to participate in a national study which could have a major impact on heart disease in our society.

Why Get Involved?

This research program is a major effort aimed directly at preventing recurrent heart attack. Your participation will help make it work. You can personally benefit from this study and future generations may also greatly benefit from your involvement.

If you are interested, or know someone else who might be interested, call your AMIS Clinic. We can answer your questions and determine if you qualify. Please do it today. The telephone number is listed on the back of this brochure.

Figure 13-1 Text of brochure used to inform potential enrollees about the Aspirin Myocardial Infarction Study. DHEW Publication No. (NIH) 76-972.

him about his family, hobbies and work so that in subsequent visits the investigator can show interest and involvement.

Clinic staff may *remind* the subject of upcoming clinic visits or study procedures.[14] Sending out postcards or calling a few days before a scheduled visit can help. A phone call has the obvious advantage that immediate feedback is obtained and a visit can be rescheduled if necessary, thereby reducing the number of subjects who fail to keep appointments. Telephoning also helps to identify a subject who is ambivalent regarding his continued participation. To preclude the clinic staff's imposing on a subject, it helps to ask in advance if the subject objects to being called frequently. Asking a subject about the best time to contact him is usually appreciated. Reminders can then be adjusted to his particular situation. In cases where subjects are reluctant to come to clinics, more than one staff person might contact the subject. For example, the physician investigator could have more influence with the subject than the staff member who usually schedules visits. In summary, the degree of interaction between an investigator and the subject seems to be important for compliance.

Many investigators *involve the subject's spouse* or other family member in the study. Informed family members can be effective supporters of the study.[14,15] They frequently make sure that visits are kept and drugs are taken on schedule. In addition, staff who are trying to alter the diet of a subject find it useful to have good rapport with the person who does the cooking. In the Multiple Risk Factor Intervention Trial,[16] which tried to reduce cholesterol by dietary means, many clinics conducted sessions during which the families of the subjects could discuss meal planning and learn new recipes.

Studies of the effects of *educating* subjects about the intervention have yielded mixed results.[11,12] Some show positive effects and others show no effect. Because of the possibility of improving compliance, with little chance of reducing it, educating the subject concerning his disease is suggested. Subjects are likely to follow dietary or exercise regimens better when they understand their relation to the disease and the rationale behind the intervention. Patient brochures, newsletters with articles addressing these issues, and group sessions for all study participants and spouses can be worthwhile, particularly in long-term trials.

Many investigators try to maintain *continuity of care*. They believe that, whenever possible, the same investigator and other staff should see the subject throughout the length of the study. While this common-sense approach should be tried, it has not been conclusively shown[17] to have made a significant impact on compliance.

Making *clinic visits pleasant* is common sense. Providing such things as parking facilities, free transportation, and comfortable waiting room facilities and minimizing waiting time will make the subject more willing to come.[11,18] Subjects having problems getting to clinic during usual working hours, might find evening or weekend sessions helpful. Home visits by staff can also be attempted. For subjects who have moved, the investigator can arrange for follow-up with physicians in other cities.

For drug studies, special *pill dispensers* help the subject keep track of when he has taken his medication.[7] Special reminders such as noticeable stickers in the bathroom or the refrigerator door or on watches have been used. Placing the pill bottles on the kitchen table or nightstand are other suggestions from subjects.

A special brochure which contains essential information and reminders may be helpful in maintaining good subject compliance (Figure 13-2). The phone number where the investigator or staff can be reached should be included in the brochure.

> **1. Your Participation in the Aspirin Myocardial Infarction Study (AMIS) is Appreciated!** AMIS, a collaborative study supported by the National Heart and Lung Institute, is being undertaken at thirty Clinics throughout the United States and involves over 4000 volunteers. As you know, this study is trying to determine whether aspirin will decrease the risk of recurrent heart attacks. It is hoped that you will personally benefit from your participating in the study and that many other people with coronary heart disease may also greatly benefit from your contribution.
>
> **2. Your Full Cooperation is Very Important to the Study.** We hope that you will follow all Clinic recommendations contained in this brochure, so that working together, we may obtain the most accurate results. If anything is not clear, please ask your AMIS Clinic Physician or Coordinator to clarify it for you. *Do not hesitate to ask questions.*
>
> **3. Keep Appointments.** The periodic follow-up examinations are very important. If you are not able to keep a scheduled appointment, call the Clinic Coordinator as soon as possible and make a new appointment. It is also important that the dietary instructions you have received, be followed carefully on the day that blood samples are drawn. At the annual visit you must be *fasting*. At the non-annual visits you are allowed to have a *fat-free diet*. Follow the directions on you Dietary Instruction Sheet. *Don't forget to take your study medication as usual on the day of your visit.*
>
> **4. Change in Residence.** If you are moving within the Clinic area, please let the Clinic Coordinator know of your change of address and telephone number as soon as possible. If you are moving away from the Clinic area, every effort will be made to arrange for continued follow-up here or at another participating AMIS Clinic.
>
> **Long Vacations.** If you are planning to leave your Clinic area for an extended period of time, let the Clinic Coordinator know so that you can be provided with sufficient study medication. Also give the Clinic Coordinator your address and telephone number so that you can be reached if necessary.
>
> **5. New Drugs.** During your participation in AMIS you have agreed not to use non-study prescribed aspirin or aspirin-containing drugs. Therefore, please call the Clinic Coordinator before starting any new drug as it might interfere with study results. At least 400 drugs contain aspirin, among them cold and cough medicines, pain relievers, ointments and salves, as well as many prescribed drugs. Many of these medications may not be labeled as to whether or not they contain aspirin or aspirin-related components. To be sure, give the Clinic Coordinator a call.

Figure 13-2 Text of brochure used to promote subject compliance in the Aspirin Myocardial Infarction Study. DHEW Publication No. (NIH) 76-1080.

---

6. **Aspirin-Free Medication**. Your Clinic will give you aspirin-free medication for headaches, other pains and fever at no cost. The following two types may be provided:
- Acetaminophen. The effects of this drug on headaches, pain and fever resemble those of aspirin. The recommended dose is 1-2 tablets every 6 hours as needed or as recommended by your Clinic Physician.
- Propoxyphene hydrochloride. The drug has an aspirin-like effect on pain only and cannot be used for the control of fever. The recommended dose is 1-2 capsules every 6 hours as needed or as recommended by your Clinic Physician.

7. **Study Medication**. You will be receiving study medication from your Clinic. You are to take two capsules each day unless prescribed otherwise. Should you forget to take your morning capsule, take it later during the day. Should you forget the evening dose, you can take it at bedtime with a glass of water or milk. The general rule is: *Do not take more than 2 capsules a day.*

8. **Under Certain Circumstances It Will Be Necessary to Stop Taking the Study Medication:**
- If you are hospitalized, stop taking the medication for the period of time you are in the hospital. Let the Clinic Coordinator know. After you leave the hospital, a schedule will be estalished for resuming medication, if it is appropriate to do so.
- If you are scheduled for surgery, we recommend that you stop taking your study medication 7 days prior to the day of the operation. This is because aspirin may, on rare occasions, lead to increased bleeding during surgery. In case you learn of the plans for surgery less than 7 days before it is scheduled, we recommend that you stop the study medication as soon as possible. And again please let the Clinic Coordinator know. After you leave the hospital, a schedule will be established for resuming medication, if it is appropriate to do so.
- If you are prescribed non-study aspirin or drugs containing aspirin by your private physician, stop taking the study medication. Study medication will be resumed when these drugs are discontinued. Let the Ciinic Coordinator know.
- If you are prescribed anti-coagulants (blood thinners), discontinue study medication and let your Clinic Coordinator know.
- If you have any adverse side effects which you think might be due to the study medication, stop taking it and call the Clinic Coordinator immediately.

9. **Study-Related Problems or Questions**. Should you, your spouse, or anyone in your family have any questions about your participation in AMIS, your Clinic will be happy to answer them. The Clinic would like you or anyone in your family to call if you have any side effects that you suspect are caused by your study medication and also if there is any change in your medical status, for example, should you be hospitalized.

10. **Your Clinic Phone Number is on the Back of This Brochure. Please Keep This Brochure as a Reference Until the End of the Study.**

Figure 13-2 (*continued*)

Contracting has received attention as a means of increasing compliance to therapeutic regimens.[19,20] However, the use of contracts between subjects and investigators in a clinical trial has not been widely implemented. The benefits of this technique in the clinical trial setting are unclear.

COMPLIANCE MONITORING

Monitoring compliance is critical, as the interpretation of study results will be influenced by knowledge of compliance with the intervention. To the extent that the control group is not truly a control and the intervention group is not being treated as intended, group differences may be diluted, leading possibly to an underestimate of the therapeutic effect and an underreporting of adverse effects. Feinstein[21] points out that differential compliance to two equally effective regimens can also lead to possibly erroneous conclusions about the effects of the intervention.

In some studies, measuring compliance is relatively easy. This is true for trials in which one group receives surgery and the other group does not, or for trials which require only a one-time intervention. Most of the time, however, assessment of compliance is not so simple. None of the measures of compliance gives a complete picture, and all are subject to possible inaccuracies and varying interpretations. Furthermore, there is no widely accepted definition or criterion for either good or poor compliance.[22]

In monitoring compliance for a long-term trial, the investigator may also be interested in changes over time. When reductions in compliance are noted, corrective action can possibly be taken. This monitoring could be by calendar time (eg, current six months versus previous six months) or by clinic visit (eg, follow-up visit number four versus previous visits). In multicenter trials, compliance with the intervention also can be looked at by clinic. In all studies, it is important for clinic staff to receive feedback about level of compliance. This can be difficult in double-blind trials where data by study group generally should not be disclosed. In this situation, the compliance data can be combined for the study groups. In non-double-blind trials, all compliance tables can be reviewed with the clinic staff.

Frequent determinations obviously have more value than infrequent ones. A better indication of true compliance can be obtained. Moreover, when the subject is aware that compliance is being monitored, frequent measures may encourage him to comply.

In drug trials, *pill or capsule count* is the easiest and most commonly used method of evaluating subject compliance. Since this assumes that the subject has ingested all medication not returned to the clinic, the validity of pill count is debated.[22,23] For example, if the subject returns the appropriate number of leftover pills at a follow-up visit, did he in fact take what he was supposed to, or did he take only some and throw out the rest? Drug dispensers can be devised which will indicate whether each dose was removed in a regular manner or all at one time.[24] However, this is expensive and cumbersome. Furthermore, taking the drug out of the dispenser does not ensure that the drug is ingested. In general, good rapport with the subjects will encourage cooperation and lead to a more accurate pill count.

Pill count is possible only as long as the pills are available to be counted. Subjects may neglect to bring their pills to the clinic to be counted. In such circumstances, the investigator may ask the subject to count the pills himself at home and to notify the investigator of the result by telephone. Obviously, less reliance can be placed on these data. How often remaining pills are not returned is important to determine. This gives an estimate of the reliability of pill count as a compliance monitor.

In monitoring pill count, the investigator may be interested in answering several questions. What is the overall compliance to the protocol prescription? If the overall compliance with the intervention is reduced, what is the main reason for the reduction? Did the investigator prescribe a reduced dose of the medication, or did the subject not follow the investigator's prescription? Is there any difference between the study groups with regard to overall compliance with protocol dosage, investigator prescription, or subject compliance with the prescribed dosage? What is the reason for reduced subject compliance? Is it because of specific side effects or is it simply forgetfulness? The answers to these questions may increase the understanding of the results of the trial.

When discussing compliance assessed by pill count, the investigator has to keep in mind that these data may be inflated and misleading, because they obviously do not include information from subjects who omit a visit. Those who miss one or more visits are often poor compliers. Therefore, the compliance data should be viewed within the framework of all subjects who should have been seen at a particular visit. Tables 13-1 and 13-2 address several of the questions raised above and illustrate ways of presenting compliance data.

Laboratory determinations can also sometimes be used to monitor compliance to medications. Tests done on either blood or urine can

Table 13-1
Average Number of Tablets Prescribed, Average Compliance to Prescription, and Average Compliance to Protocol, by Clinic, Study Groups Combined*

Clinic	Average No. of Tablets Prescribed Per Day	Average No. of Tablets Taken Per Day	Average Compliance to Prescription†	Average Compliance to Protocol‡
A	1.97	1.92	97.5%	96.0%
B	1.83	1.67	91.3%	83.5%
C	1.96	1.88	95.9%	94.0%
Total	1.92	1.82	94.9%	91.0%

*Table used in Aspirin Myocardial Infarction Study: Coordinating Center, University of Maryland.
†Average compliance to prescription is average number of tablets taken per day divided by average number of tablets prescribed per day.
‡Average compliance to protocol is average number of tablets taken per day divided by daily protocol dosage (in this case, two tablets per day).

detect the presence of active drugs or metabolites. A limitation in measuring substances in urine or blood is the short half-life of most drugs. Therefore, laboratory determinations usually indicate only what has happened in the preceding day or two. A control subject who takes the active drug (obtained from a source outside the trial) until the day prior to a clinic visit, or a subject in the intervention group who takes the active drug only on the day of the visit might not be detected as being a poor complier. Moreover, drug compliance in subjects taking an inert placebo tablet cannot be assessed by any laboratory determination. Adding a specific chemical substance such as riboflavin can serve as a marker in cases where the drug or its metabolites are difficult to measure.[25] However, the same drawbacks apply to markers as to masking substances—the risk of toxicity in long-term use may outweigh benefits. Use of markers is discouraged unless a marker is clearly established as both safe and unlikely to confound assessment of response variables.

Tables 13-3 and 13-4 illustrate how compliance based on a laboratory determination can be monitored. The examples are from a placebo-control trial of aspirin[26] in which the salicylate level in urine was used as an indicator of drug compliance. The tables were designed to provide information on dropouts (subjects in the aspirin group not taking aspirin) and dropins (subjects in the placebo group taking aspirin). Ideally, the percentages in Table 13-3 ought to be 100 and 0 for Group A and Group B, respectively, but they very rarely are. The investigator should determine in advance, preferably in a pre-test, the definition of an indicator of positive compliance. The selection of the cut-off can make compliance look good or bad. If the investigator chooses a high level of urine salicylate and accepts only values over that cut-off as being

Table 13-2
Average Number of Tablets Prescribed, Average Compliance to Prescription, and Average Compliance to Protocol, by Follow-up Visit and Study Group

Visit	Average No. of Tablets Prescribed Per Day		Average No. of Tablets Taken Per Day		Average Compliance to Prescription*		Average Compliance to Protocol†	
	Group A	Group B	Group A	Group B	Group A	Group B	Group A	Group B
1	1.95	1.94	1.91	1.89	97.9%	97.4%	95.5%	94.5%
2	1.93	1.90	1.82	1.78	94.3%	93.7%	91.0%	89.0%
3	1.92	1.89	1.83	1.80	95.3%	95.2%	91.5%	90.0%

*Average compliance to prescription is average number of tablets taken per day divided by average number of tablets prescribed per day.
†Average compliance to protocol is average number of tablets taken per day divided by daily protocol dosage (in this case, two tablets per day).

positive, he will probably have a high dropout rate, because a proportion of subjects who faithfully take aspirin may have low levels of urine salicylate. The interval between the time the last tablet was taken and the time the urine specimen was obtained might be long enough for most of the salicylate to be excreted. When the cut-off level is low, the investigator runs the risk of finding what appears to be a high dropin rate. In fact, these dropins can be attributable to imprecision of the method at low values. In addition, there may be false positive results. For example, a urine salicylate determination is not specific for aspirin. Products which contain methyl salicylate can also give measurable levels of salicylate in the urine. Again, caution is necessary in a double-blind trial when compliance tables based on laboratory determinations of the active drug or metabolite are shared with the clinic staff. Combining data from the study groups as shown in Table 13-4 is one way of overcoming the problem of unblinding.

Table 13-3
Percent of Subjects With Positive Urine Salicylate
by Visit and Study Group

	Group A		Group B	
Visit	Subjects Completing Visits	Percent of Subjects with Positive Test*	Subjects Completing Visits	Percent of Subjects with Positive Test*
1	260	88.5	264	3.0
2	251	86.1	250	4.4
3	242	86.8	237	4.2

*Urine salicylate $\geq$ a mmol/L.

Table 13-4
Percent of Subjects Complying to Protocol* as Defined by Urine
Salicylate Tests, by Clinic and Follow-up Visit, Study Groups Combined

	Follow-up Visit 1		Follow-up Visit 2	
Clinic	Subjects Completing Visits	Percent Compliers	Subjects Completing Visits	Percent Compliers
A	173	92.5	165	90.9
B	145	90.3	141	87.9
C	206	94.7	195	92.8
Total	524	92.7	501	90.8

*Compliance is defined as:
 a. A subject randomized to Group A (aspirin) having a urine salicylate level $\geq$ a mmol/L.
 b. A subject randomized to Group B (placebo) having a urine salicylate level $\leq$ b mmol/L.

Laboratory tests obtained on occasions not associated with clinic visits might give a better picture of compliance. Thus, the subject could be instructed, at certain intervals, to send to the clinic a vial of urine. Such a technique is of value only so long as the subject does not associate it with a compliance monitoring procedure. In at least one study,[26] information obtained in this manner contributed no additional information to laboratory results done at scheduled visits, except perhaps as a confirmation of such results.

Measurement of *physiological response variables* can be helpful. Cholesterol reduction by drug or diet is unlikely to occur in one or two days. Therefore, a subject in the intervention group cannot suddenly comply with the regimen the day before a clinic visit and expect to go undetected. Similarly, a control subject's cholesterol is unlikely to rise in the one day before a visit that he was off active drug. Other physiological response variables that might be monitored are blood pressure in an antihypertensive study, carbon monoxide in a smoking study, platelet aggregation in an aspirin study, and graded exercise in an exercise study. In all these cases, the indicated response variable would not be the primary response variable but merely an intermediate indicator of compliance to the intervention regimen. Unfortunately, other measures, such as triglyceride levels, are highly variable; indications of noncompliance of individual subjects using these measures are not easily interpreted. Group data, however, may be useful.

Other compliance monitoring techniques fall under the category of *interview* or *record keeping*. A diet study might use a 24-hour recall or a 7-day food record. Exercise studies might use diaries or charts to indicate frequency and kind of exercise. Studies of people with angina pectoris might record frequency of attacks of pain and of nitroglycerine administration. The techniques mentioned depend greatly on subject recall (even diaries require subjects to remember to make entries) and are subjective. As a result, interpretations of compliance in individual subjects can be misleading. Group data, again, are more reliable. Interviews of the subjects have been proposed as a useful method of measuring subject compliance with the intervention.[27] However, others[28] feel that an interview alone is an unreliable indicator. Most subjects tend to overestimate their compliance either in an effort to please the investigator or because of faulty memory.[22] In the Multiple Risk Factor Intervention Trial, the investigators measured serum thiocyanate to adjust for the subjects' claims of quitting smoking.[29] As seen in Table 13-5, a substantial correction was made in the Special Intervention group, particularly at year 1.

Investigator ratings of compliance have been utilized.[22] It is probably the least reliable of the various techniques, seriously overestimating compliance.[22,28]

The kinds of compliance monitoring and the uses to which they can

Table 13-5
Reported and Thiocyanate-Adjusted Cigarette Smoking Quit Rate (%) by Group in the Multiple Risk Factor Intervention Trial[24]

	Special Intervention		Usual Care	
Year	*Reported*	*Adjusted*	*Reported*	*Adjusted*
1	43	31	14	12
6	50	46	29	29

be put depend on whether the study is blinded. In a double-blind study, laboratory measurements and physiological response variables on individual subjects must be kept from the examining investigator. They can be used only in evaluating total study results and not in ongoing maintenance of good compliance. For this purpose, the investigator must rely on pill count. In single-blind and unblinded studies, all of the techniques mentioned above can be used to encourage good subject compliance.

Another aspect of monitoring deals with subject compliance to study procedures such as attendance at scheduled visits. One of the purposes of these visits is to collect response variable data. The data will be better if they are more complete. Thus, by itself, completeness of data can be a measure of the quality of a clinical trial. Studies with even a moderate amount of missing data or subjects lost to follow-up could give misleading results and should be interpreted with caution. By reviewing the reasons why subjects missed scheduled clinic visits, the investigator can identify factors that can be corrected or improved. Having the subjects come in for study visits facilitates and encourages compliance to study medication. Study drugs are dispensed at these visits and the dose is adjusted when necessary. Tables 13-6 and 13-7 show ways of monitoring missed visits.

Table 13-6
Missed Visits by Study Group and Follow-up Visit

	Group A			Group B		
Visit	*Possible Visits*	*No. Missed Visits*	*Percent Missed Visits*	*Possible Visits*	*No. Missed Visits*	*Percent Missed Visits*
1	270	10	3.7	272	8	2.9
2	263	12	4.6	258	8	3.1
3	257	15	5.8	251	14	5.6
Total	790	37	4.7	781	30	3.8

Table 13-7
Missed Visits by Clinic and Follow-up Visit, Study Groups Combined

Clinic	Follow-up Visit 1			Follow-up Visit 2		
	Possible Visits	No. Missed Visits	Percent Missed Visits	Possible Visits	No. Missed Visits	Percent Missed Visits
A	177	4	2.3	170	5	2.9
B	153	8	5.2	148	7	4.7
C	212	6	2.8	203	8	3.9
Total	542	18	3.3	521	20	3.8

From a statistical viewpoint every randomized subject should be included in the analysis (Chapters 5 and 16). Consequently, the investigator must keep trying to get all subjects back for scheduled visits until the trial is over. Even if a subject is taken off his study medication by an investigator or if he stops taking it on his own, he should be encouraged to come in for his regular study visits. This enables the investigator to get more complete follow-up data. In addition, subjects do change their minds. For a long time, they may want to have nothing to do with the trial and later may agree to come back for visits and even resume taking their assigned intervention regimen. A program for returning dropouts has been reported.[30] Special attention to each dropout's problems and an emphasis on potential contribution to the trial led to successful retrieval of a large proportion. Inasmuch as the subject will be counted in the analysis, leaving open the option for the subject to return to active participation in the study is worthwhile.

REFERENCES

1. CASS Principal Investigators and Their Associates. Coronary Artery Surgery Study (CASS): a randomized trial of coronary artery bypass surgery. Survival data. *Circulation* 68:939–950, 1983.

2. Haynes, R.B., Taylor, D.W., and Sackett, D.L. (Eds): *Compliance in Health Care*. Baltimore: Johns Hopkins University Press, 1979.

3. Eraker, S.A., Kirscht, J.P., and Becker, M.H. Understanding and improving patient compliance. *Ann Intern Med*. 100:258–268, 1984.

4. Sackett, D.L., and Snow, J.C. The magnitude of compliance and noncompliance. Edited by R.B. Haynes, D.W. Taylor, and D.L. Sackett. In *Compliance in Health Care*. Baltimore: Johns Hopkins University Press, 1979.

5. Sanne, H. Exercise tolerance and physical training of non-selected patients after myocardial infarction. *Acta Med Scand*. (suppl 551):1–124, 1973.

6. National Diet-Heart Study Research Group. Diet Heart Study. Final report. *Circulation* 37(suppl I):I-1–I-428, 1968.

7. McKenney, J.M. The clinical pharmacy and compliance. Edited by R.B. Haynes, D.W. Taylor, and D.L. Sackett. In *Compliance in Health Care*. Baltimore: Johns Hopkins University Press, 1979.

8. Becker, M.H., Maiman, L.A., Kirscht, J.P., et al. Patient perceptions and compliance: recent studies of the health belief model. Edited by R.B. Haynes, D.W. Taylor, and D.L. Sackett. In *Compliance in Health Care*. Baltimore: Johns Hopkins University Press, 1979.

9. Shulman, N., Cútter, G., Daugherty, R., et al. Correlates of attendance and compliance in the Hypertension Detection and Follow-up Program. *Controlled Clin Trials* 3:13–27, 1982.

10. Goldman, A.I., Holcomb, R., Perry, H.M., Jr., et al. Can dropout and other noncompliance be minimized in a clinical trial? Report from the Veterans Administration—National Heart, Lung, and Blood Institute cooperative study on antihypertensive therapy: mild hypertension. *Controlled Clin Trials* 3:75–89, 1982.

11. Haynes, R.B. Strategies to improve compliance with referrals, appointments, and prescribed medical regimens. Edited by R. B. Haynes, D.W. Taylor, and D.L. Sackett. In *Compliance in Health Care*. Baltimore: Johns Hopkins University Press, 1979.

12. Green, L.W. Educational strategies to improve compliance with therapeutic and preventive regimens: the recent evidence. Edited by R.B. Haynes, D.W. Taylor, and D.L. Sackett. In *Compliance in Health Care*. Baltimore: Johns Hopkins University Press, 1979.

13. Buring, J., and Hennekens, C.H. Sample size and compliance in randomized trials. Edited by M.A. Sestili. In *Chemoprevention Clinical Trials: Problems and Solutions*. NIH Publication No. 85-2715, 1984.

14. Dunbar, J.M., Marshall, G.D., and Hovell, M.F. Behavioral strategies for improving compliance. Edited by R.B. Haynes, D.W. Taylor, and D.L. Sackett. In *Compliance in Health Care*. Baltimore: Johns Hopkins University Press, 1979.

15. Hogue, C.C. Nursing and compliance. Edited by R.B. Haynes, D.W. Taylor, and D.L. Sackett. In *Compliance in Health Care*. Baltimore: Johns Hopkins University Press, 1979.

16. Mandriota, R., Bunkers, E.B., and Wilcox, M.E. Nutrition intervention strategies in MRFIT. Presented at the 61st Annual Meeting of the American Dietetic Association, New Orleans, September 25–29, 1978.

17. Hulka, B.S. Patient-clinician interactions and compliance. Edited by R.B. Haynes, D.W. Taylor, and D.L. Sackett. In *Compliance in Health Care*. Baltimore: Johns Hopkins University Press, 1979.

18. Haynes, R.B. Determinants of compliance: the disease and the mechanics of treatment. Edited by R.B. Haynes, D.W. Taylor, and D.L. Sackett. In *Compliance in Health Care*. Baltimore: Johns Hopkins University Press, 1979.

19. Vance, B. Using contracts to control weight and to improve cardiovascular physical fitness. Edited by J.D. Krumboltz and C.E. Thoresen. In *Counseling Methods*. New York: Holt, Rinehart, and Winston, 1976.

20. Steckel, S.B., and Swain, M.A. Contracting with patients to improve compliance. *Hospitals* 51:81–84, 1977.

21. Feinstein, A. Clinical biostatistics. XXX. Biostatistical problems in "compliance bias." *Clin Pharmacol Ther*. 16:846–857, 1974.

22. Dunbar, J. Assessment of medication compliance: A review. Edited by R.B. Haynes, M.E. Mattson, and T.O. Engebretson, Jr. In *Patient Compliance to Prescribed Antihypertensive Medication Regimens: A Report to the National Heart, Lung, and Blood Institute*. DHHS Pub. No. (NIH) 81-2102.

23. Gordis, L. Conceptual and methodologic problems in measuring patient compliance. Edited by R.B. Haynes, D.W. Taylor, and D.L. Sackett. In *Compliance in Health Care*. Baltimore: Johns Hopkins University Press, 1979.

24. Moulding, T.S. The unrealized potential of the medication monitor. *Clin Pharmacol Ther*. 25:131–136, 1979.

25. Workshop on the development of markers to use as adherence measures in clinical studies. Edited by T. Blaszkowski and W. Insull, Jr. *Controlled Clin Trials* 5:451–588, 1984.

26. Aspirin Myocardial Infarction Study Research Group. A randomized, controlled trial of aspirin in persons recovered from myocardial infarction, *JAMA* 243:661–669, 1980.

27. Fletcher, S.W., Pappius, E.M., and Harper, S.J. Measurement of medication compliance in a clinical setting. *Arch Intern Med*. 139:635–638, 1979.

28. Roth, H.P., and Caron, H.S. Accuracy of doctors' estimates and patients' statements on adherence to a drug regimen. *Clin Pharmacol Ther*. 23:361–370, 1978.

29. Multiple Risk Factor Intervention Trial Research Group. Multiple Risk Factor Intervention Trial. Risk factor changes and mortality results. *JAMA* 248:1465–1477, 1982.

30. Probstfield, J.L., Russell, M.L., Henske, J.C., et al. A successful program for returning dropouts to a clinical trial. *Controlled Clin Trials*. 1:168, 1980.

CHAPTER 14

Survival Analysis

This chapter reviews some of the fundamental concepts and basic methods in survival analysis. Frequently, event rates such as mortality or frequency of nonfatal myocardial infarction are selected as primary response variables. The analysis of such event rates in two groups could employ the chi-square statistic or the equivalent normal statistic for the comparison of two proportions. However, since the length of observation is often different for each subject, estimating an event rate is more complicated. Furthermore, simple comparison of event rates between two groups is not necessarily the most informative type of analysis. For example, the five-year survival for two groups may be nearly identical, but the survival rates may be quite different at various times during the five years. This is illustrated by the survival curves in Figure 14-1. This figure shows survival probability on the vertical axis and time on the horizontal axis. For Group *A,* the survival rate (or one minus the mortality rate) declines steadily over the five years of observation. For Group *B,* however, the decline in the survival rate is rapid during the first year and then levels off. Obviously, the survival experience of the two groups is not the same, although the mortality rate at five years is nearly the same. If only the five-year survival rate—instead of the five-year survival experience—is considered, Group *A* and Group *B* appear equivalent. Curves such as these might reasonably be expected in a trial of surgical versus medical intervention, where surgery might carry a high initial operative mortality.

FUNDAMENTAL POINT

Survival analysis methods are important in trials where subjects are entered over a period of time and have various lengths of follow-up. These methods permit the comparison of the entire survival experience during the follow-up and may be used for the analysis of time to any dichotomous response variable such as a nonfatal event or an adverse effect.

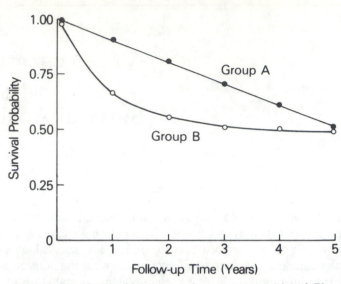

Figure 14-1 Survival experience for two groups (*A* and *B*).

Some elementary statistical textbooks[1,2] review basic techniques of survival analysis. A more complete review is in other texts.[3-5] Many methodological advances in the field have occurred in recent years and this book will not be able to cover all developments. The following discussion will concern two basic aspects: first, estimation of the survival experience or survival curve for a group of subjects in a clinical trial and second, comparison of two survival curves to test whether the survival experience is significantly different. Although the term survival analysis is used, the methods are more widely applicable than to just survival. The methods can be used for any dichotomous response variable in a clinical trial when the time from enrollment to the time of the event — not just the fact of its occurrence — is an important consideration.

ESTIMATION OF THE SURVIVAL CURVE

The graphical presentation of the total survival experience during the period of observation is called the survival curve, and the tabular presentation is called the lifetable. This chapter will discuss two similar methods, the Kaplan-Meier method[6] and the Cutler-Ederer method[7] for estimating the true survival curve or lifetable. Before the review of these specific methods, however, it is necessary to explain how survival experience is typically obtained in a clinical trial and to define some of the associated terminology.

The clinical trial design may, in a simple case, require that all subjects be observed for T years. This is referred to as the follow-up time or exposure time. If all the subjects are entered as a single cohort at the same time, the actual period of follow-up is the same for all subjects. If, however, as in most clinical trials, the entry of subjects is staggered over some recruitment period, then the T year period of follow-up may be a different actual or calendar time for each subject, as illustrated in Figure 14-2.

During the course of follow-up, a subject may have an event. The event time is the accumulated time from entry into the study to the event. The interest is not in the actual calendar date when the event took place but rather the interval of time from entry into the trial until the event. Figures 14-3 and 14-4 illustrate the way the actual survival experience for staggered entry of subjects is translated for the analysis. In Figure 14-3, subjects 2 and 4 have an event while subjects 1 and 3 do not during the follow-up time. Since, for each subject, only the time interval from entry to the end of the scheduled follow-up period or until an event is of interest, the time of entry can be considered as time zero for each subject. Figure 14-4 illustrates the same survival experience as Figure 14-3, but the time of entry is considered as time zero.

In most clinical trials, the investigator may be unable to assess the occurrence of the event in some subjects. The follow-up time or exposure time for these subjects is said to be censored; that is, the investigator does not know what happened to these subjects after they stopped participating in the trial. Another example of censoring is when subjects are entered in a staggered fashion, and the study is terminated at a common date before all subjects have had at least their complete T years of

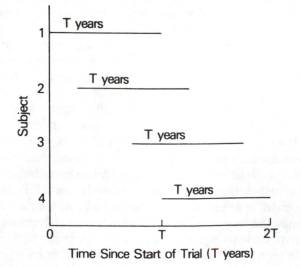

Figure 14-2 T year follow-up time for four subjects with staggered entry.

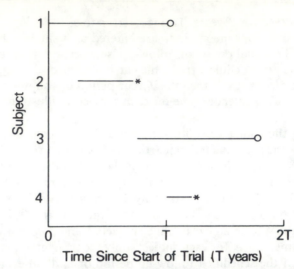

Figure 14-3 Follow-up experience of four subjects with staggered entry: two subjects with observed events (∗) and two subjects followed for time *T* without events (o).

follow-up. These subjects are also considered as losses, but the reason for censoring is administrative. Administrative censoring could also occur if a trial is terminated prior to the scheduled time because of early benefits or harmful effects of the intervention. In all cases, censoring is assumed to be independent of occurrence of events.

Figure 14-5 illustrates several of the possibilities for observations during follow-up. Note that in this example the investigator has planned to follow all subjects to a common termination time, with each subject being followed for at least *T* years. The first three subjects were randomized at the start of the study. The first subject was observed for the entire duration of the trial with no event, and his survival time was censored because of study termination. The second subject had an event before the end of follow-up. The third subject was lost to follow-up. The second group of three subjects was randomized later during the course of the trial with experiences similar to the first group of three. Subjects 7 through 11 were randomized late in the study and were not able to be followed for at least *T* years because the study was terminated early. Subject 7 was lost to follow-up and subject 8 had an event before *T* years of follow-up time had elapsed and before the study was terminated. Subject 9 was administratively censored but theoretically would have been lost to follow-up had the trial continued. Subject 10 was also censored because of early study termination, although he had an event afterwards which would have been observed had the trial continued to its scheduled end. Finally, the last subject who was censored would have survived for at

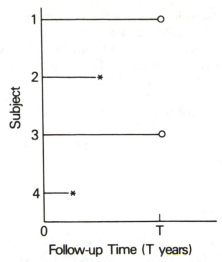

Figure 14-4 Follow-up experience of four subjects with staggered entry converted to a common starting time: two subjects with observed events (*) and two subjects followed for time *T* without events (o).

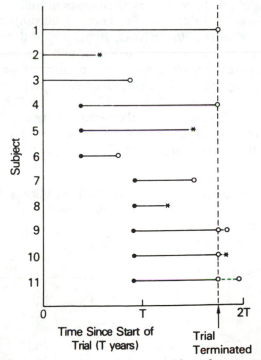

Figure 14-5 Follow-up experience of eleven subjects for staggered entry and a common termination time, with observed events (*) and censoring (o). Follow-up experience beyond the termination time is shown for subjects 9 through 11.

least T years had the study lasted as long as first planned. The survival experiences illustrated in Figure 14-5 would all be shifted to have a common starting time equal to zero as in Figure 14-4. The follow-up time, or the time elapsed from calendar time of entry to calendar time of an event or to censoring could then be analyzed.

In summary then, the investigator needs to record for each subject the time of entry and the time of an event, the time of loss to follow-up, or whether the subject was still being followed without having had an event when the study is terminated. These data will allow him to compute the survival curve.

Kaplan-Meier Estimate

In a clinical trial with staggered entry of subjects and censored observations, survival data will be of varying degrees of completeness. As a very simple example, suppose that 100 subjects were entered into a study and followed for two years. One year after the first group was started, a second group of 100 subjects was entered and followed for the remaining year of the study. Assuming no losses to follow-up, the results might be as shown in Table 14-1. For Group I, 20 subjects died during the first year and of the 80 survivors, 20 more died during the second year. For Group II, which was followed for only one year, 25 subjects died. Now suppose the investigator wants to estimate the two-year survival rate. The only group of subjects followed for two years was Group I. One estimate of two-year survival, $P(2)$, would be $P(2) = 60/100$ or 0.60. Note that the first-year survival experience of Group II is ignored in this estimate. If the investigator wants to estimate one year survival rate, $P(1)$, he would observe that a total of 200 subjects were followed for

Table 14-1
Subjects Entered at Two Points in Time (Group I and Group II) and Followed to a Common Termination Time*

Years of Follow-up		Group	
		I	II
1	Subjects Entered	100	100
	1st Year Deaths	20	25
	1st Year Survivors	80	75
2	Subjects Entered	80	
	2nd Year Deaths	20	
	2nd Year Survivors	60	

*After Kaplan and Meier.[6]

at least one year. Of those, 155 (80 + 75) survived the first year. Thus, $P(1) = 155/200$ or 0.775. If each group were evaluated separately, the survival rates would be 0.80 and 0.75. In estimating the one-year survival rate, all the available information was used, but for the two-year survival rate the one-year survival experience of Group *II* was ignored.

Another procedure for estimating survival rates is to use a conditional probability. For this example, the probability of two-year survival, $P(2)$, is equal to the probability of one-year survival, $P(1)$, times the probability of surviving the second year, given that the subject survived the first year, $P(2|1)$. That is, $P(2) = P(1) P(2|1)$. In this example, $P(1) = 0.775$. The estimate for $P(2|1)$ is $60/80 = 0.75$ since 60 of the 80 subjects who survived the first year also survived the second year. Thus, the estimate for $P(2) = 0.775 \times 0.75$ or 0.58, which is slightly different from the previously calculated estimate of 0.60.

Kaplan and Meier[6] described how this conditional probability strategy could be used to estimate survival curves in clinical trials with censored observations. Their procedure is usually referred to as the Kaplan-Meier estimate, or sometimes the product-limit estimate, since the product of conditional probabilities leads to the survival estimate. This procedure assumes that the exact time of entry into the trial is known and that the exact time of the event or loss of follow-up is also known. For some applications, time to the nearest month may be sufficient, while for other applications the nearest day or hour may be necessary. Kaplan and Meier assumed that a death and loss of follow-up would not occur at the same time. If a death and a loss to follow-up are recorded as having occurred at the same time, this tie is broken on the assumption that the death occurred slightly before the loss to follow-up.

In this method, the follow-up period is divided into intervals of time so that no interval contains both deaths and losses. Let P_j be equal to the probability of surviving the jth interval, given that the subject has survived the previous interval. For intervals labeled j with deaths only, the estimate for P_j, which is $\hat{P}_j$, is equal to the number of subjects alive at the beginning of the jth interval, n_j, minus those who died during the interval, δ_j, with this difference being divided by the number alive at the beginning of the interval, ie, $\hat{P}_j = (n_j - \delta_j)/n_j$. For an interval j with only λ_j losses, the estimate $\hat{P}_j$ is one. Such conditional probabilities for an interval with only losses would not alter the product. This means that an interval with only losses and no deaths may be combined with the previous interval.

As a simple example, suppose 20 subjects are followed for a period of one year, and to the nearest tenth of a month, deaths were observed at the following times: 0.5, 1.5, 1.5, 3.0, 4.8, 6.2, 10.5 months. In addition, losses to follow-up were recorded at: 0.6, 2.0, 3.5, 4.0, 8.5, 9.0 months. It is convenient for illustrative purposes to list the deaths and losses together in ascending time with the losses indicated in parentheses. Thus, the following sequence is obtained: 0.5, (0.6), 1.5, 1.5, (2.0), 3.0, (3.5), (4.0), 4.8, 6.2, (8.5), (9.0), 10.5. The remaining seven subjects were all censored at 12 months due to termination of the study.

Table 14-2 presents the survival experience for this example as a lifetable. Each row in the lifetable indicates the time at which a death or an event occurred. One or more deaths may have occurred at the same time and they are included in the same row in the lifetable. In the interval between two consecutive times of death, losses to follow-up may have occurred. Hence, a row in the table actually represents an interval of time, beginning with the time of a death—up to but not including the time of the next death. In this case, the first interval is defined by the death at 0.5 months up to the time of the next death at 1.5 months. The columns labeled n_j, δ_j, and λ_j correspond to the definitions given above and contain the information from the example. In the first interval, all 20 subjects were initially at risk, one died at 0.5 months, and later in the interval (at 0.6 months) one subject was lost to follow-up. In the second interval, from 1.5 months up to 3.0 months, 18 subjects were still at risk initially, two deaths were recorded at 1.5 months and one subject was lost at 2.0 months. The remaining intervals are defined similarly. The column labeled $\hat{P}_j$ is the conditional probability of surviving the interval j and is computed as $(n_j - \delta_j)/n_j$ or $(20 - 1)/20 = 0.95$, $(18 - 2)/18 = 0.89$, etc. The column labeled $\hat{P}(t)$ is the estimated survival curve and is computed as the accumulated product of the $\hat{P}_j$'s ($0.84 = 0.95 \times 0.89$, $0.79 = 0.95 \times 0.89 \times 0.93$, etc).

Table 14-2
Kaplan-Meier Lifetable for 20 Subjects Followed for One Year

Interval	Interval Number	Time of Death	n_j	δ_j	λ_j	$\hat{P}_j$	$\hat{P}(t)$	$V[\hat{P}(t)]$
[0.5,1.5)	1	0.5	20	1	1	0.95	0.95	0.0024
[1.5,3.0)	2	1.5	18	2	1	0.89	0.84	0.0068
[3.0,4.8)	3	3.0	15	1	2	0.93	0.79	0.0089
[4.8,6.2)	4	4.8	12	1	0	0.92	0.72	0.0114
[6.2,10.5)	5	6.2	11	1	2	0.91	0.66	0.0133
[10.5,)	6	10.5	8	1	7*	0.88	0.57	0.0161

n_j: number of subjects alive at the beginning of the jth interval
δ_j: number of subjects who died during the jth interval
λ_j: number of subjects who were lost or censored during the jth interval
$\hat{P}_j$: estimate for P_j, the probability of surviving the jth interval given that the subject has survived the previous intervals
$\hat{P}(t)$: estimated survival curve
$V[\hat{P}(t)]$: variance of $\hat{P}(t)$
*Censored due to termination of study.

The graphical display of the next to last column of Table 14-2, $\hat{P}(t)$, is given in Figure 14-6. The step function appearance of the graph is because the estimate of $P(t)$, $\hat{P}(t)$, is constant during an interval and changes only at the time of a death. With very large sample sizes and more observed deaths, the step function has smaller steps and looks more like the usually visualized smooth survival curve. If no censoring occurs, this method simplifies to the number of survivors divided by the total number of subjects who entered the trial.

Since $\hat{P}(t)$ is an estimate of $P(t)$, the true survival curve, the estimate will have some variation due to the sample selected. Greenwood[8] derived a formula for estimating the variance of an estimated survival function which is applicable to the Kaplan-Meier method. The formula for the variance of $\hat{P}(t)$, denoted $V[\hat{P}(t)]$ is given by

$$V[\hat{P}(t)] = \hat{P}^2(t) \sum_{j=1}^{K} \frac{\delta_j}{n_j(n_j - \delta_j)}$$

where n_j and δ_j are defined as before, and $K =$ the number of intervals. In Table 14-2, the last column labeled $V[\hat{P}(t)]$ represents the estimated variances for the estimates of $P(t)$ during the 6 intervals. Note that the variance increases as one moves down the column. When fewer subjects

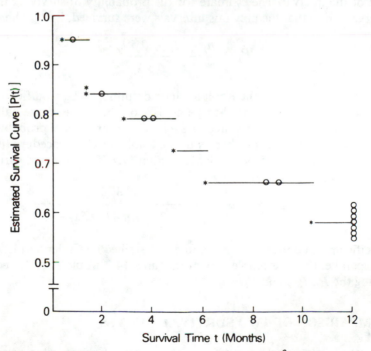

Figure 14-6 Kaplan-Meier estimate of a survival curve, $\hat{P}(t)$, from a one-year study of 20 subjects, with observed events (*) and censoring (o).

are at risk, the ability to estimate the survival experience is diminished.

Other examples of this procedure, as well as a more detailed discussion of some of the statistical properties of this estimate, are provided by Kaplan and Meier.[6] Computer programs are available[9] so that survival curves can be obtained quickly, even for very large sets of data.

Cutler-Ederer Estimate

In the Kaplan-Meier estimate, it was required that the exact time of death or loss be known so that the observations could be ranked, or at least grouped appropriately, into intervals with deaths preceding losses. For some studies, all that is known is that within an interval of time from t_{j-1} to t_j, denoted (t_{j-1}, t_j), δ_j deaths and λ_j losses occurred among the n_j subjects at risk. Within that interval the order in which the events and losses occurred is unknown. In the Kaplan-Meier procedure, the intervals were chosen so that all deaths preceded all losses in any interval.

In the Cutler-Ederer or actuarial estimate,[7] the assumption is made that the deaths and losses are uniformly distributed over an interval. On the average, this means that one half the losses will occur during the first half of the interval. The estimate for the probability of surviving the jth interval, given that the previous intervals were survived, is $\hat{P}_j$, where

$$\hat{P}_j = \frac{n_j - \delta_j - 0.5\,\lambda_j}{n_j - 0.5\,\lambda_j}$$

Notice the similarity to the Kaplan-Meier definition. The modification is that the λ_j losses are assumed to be at risk, on the average, one half the time and thus should be counted as such. These conditional probabilities $\hat{P}_j$ are then multiplied together as in the Kaplan-Meier procedure to obtain an estimate $\hat{P}(t)$ of the survival function $P(t)$. The estimated variance for $\hat{P}(t)$ in this case is given by

$$V[\hat{P}(t)] = \hat{P}^2(t) \sum_{j=1}^{K} \frac{\delta_j}{(n_j - 0.5\lambda_j)(n_j - 0.5\lambda_j - \delta_j)}$$

Specific applications of this method are described by Cutler and Ederer. The parallel to the example shown in Table 14-2 would require recomputing the $\hat{P}_j$, $\hat{P}(t)$ and $V[\hat{P}(t)]$.

COMPARISON OF TWO SURVIVAL CURVES

We have just discussed how to estimate the survival curve in a clinical trial for a single group. For two groups, the survival curve would be estimated for each group separately. The question is whether the two survival curves $P_C(t)$ and $P_I(t)$, for the control and intervention groups respectively, are different based on the estimates $\hat{P}_C(t)$ and $\hat{P}_I(t)$.

Point-by-Point Comparison

One possible comparison between groups is to specify a time t^* for which survival estimates have been computed using the Kaplan-Meier or Cutler-Ederer method. At time t^*, one can compare the survival estimates $\hat{P}_C(t^*)$ and $\hat{P}_I(t^*)$ using the statistic

$$Z(t^*) = \frac{\hat{P}_C(t^*) - \hat{P}_I(t^*)}{\{V[\hat{P}_C(t^*)] + V[\hat{P}_I(t^*)]\}^{\frac{1}{2}}}$$

where $V[\hat{P}_C(t^*)]$ and $V[\hat{P}_I(t^*)]$ are the Greenwood estimates of variance.[8] The statistic $Z(t^*)$ has approximately a normal distribution with mean zero and variance one under the null hypothesis that $P_C(t^*) = P_I(t^*)$. The problem with this approach is the multiple looks issue described in Chapter 15. Another problem exists in interpretation. For example, what conclusions should be drawn if two survival curves are judged significantly different at time t^* but not at any other points? The issue then becomes, what point in the survival curve is most important.

For some studies with a T year follow-up, the T year mortality rates are considered important and should be tested in the manner just suggested. Annual rates might also be considered important and, therefore, compared. One criticism of this suggestion is that the specific points may have been selected to yield the largest difference based on the observed data. One can easily visualize two survival curves for which significant differences are found at a few points. However, when survival curves are compared, the large differences indicated by these few points are not supported by the overall survival experience. Therefore, point-by-point comparisons are not recommended unless a few points can be justified prior to data analysis and are specified in the protocol.

Total Curve Comparison

Because of the limitations of comparison of point-by-point estimates, Gehan[10] and Mantel[11] proposed statistical methods to assess the overall survival experience. These two methods were important steps in the development of analytical methods for survival data. They both assume that the hypothesis being tested is whether two survival curves are equal, or whether one is consistently different from the other. If the two survival curves cross, these methods should be interpreted cautiously.

Mantel[11] proposed the use of the procedure described by Cochran[12] and Mantel and Haenszel[13] for combining a series of 2×2 tables. In this procedure, each time, t_j, a death occurs in either group, a 2×2 table is formed as follows:

	Deaths at time t_j	Survivors at time t_j	At risk prior to time t_j
Intervention	a_j	b_j	$a_j + b_j$
Control	c_j	d_j	$c_j + d_j$
	$a_j + c_j$	$b_j + d_j$	n_j

The entry a_j represents the observed number of deaths at time t_j in the intervention group and c_j represents the observed number of deaths at time t_j in the control group. At least a_j or c_j must be non-zero. One could create a table at other time periods (that is, when a_j and c_j are zero), but this table would not make any contribution to the statistic. Of the n_j subjects at risk just prior to time t_j, $a_j + b_j$ were in the intervention group and $c_j + d_j$ were in the control group. The expected number of deaths in the intervention group, denoted $E(a_j)$, can be shown to be

$$E(a_j) = (a_j + c_j)(a_j + b_j)/n_j$$

and the variance of the observed number of deaths in the intervention group, denoted as $V(a_j)$ is given by

$$V(a_j) = \frac{(a_j + c_j)(b_j + d_j)(a_j + b_j)(c_j + d_j)}{n_j^2 (n_j - 1)}$$

These expressions are the same as those given for combining 2×2 tables in the Appendix of Chapter 16. The Mantel-Haenszel *(MH)* statistic is given by

$$MH = \{\Sigma_{j=1}^{K} a_j - E(a_j)\}^2 / \Sigma_{j=1}^{K} V(a_j)$$

and has approximately a chi-square distribution with one degree of freedom, where K is the number of distinct event times in the combined intervention and control groups.

Application of this procedure is straightforward. First, the times of deaths and losses in both groups are ranked in ascending order. Second, the time of each death, and the total number of subjects in each group who were at risk just before the death (a_j+b_j, c_j+d_j) as well as the number of deaths in each group (a_j, c_j) are determined. With this information, the appropriate 2×2 tables can be formed.

Example Assume that the data in the example shown in Table 14-2 represent the data from the control group. Among the 20 subjects in the intervention group, two deaths occurred at 1.0 and 4.5 months with losses at 1.6, 2.4, 4.2, 5.8, 7.0, and 11.0 months. The observations, with parentheses indicating losses, can be summarized as follows:

Intervention: 1.0, (1.6), (2.4), (4.2), 4.5, (5.8), (7.0), (11.0)

Control: 0.5, (0.6), 1.5, 1.5, (2.0), 3.0, (3.5), (4.0), 4.8, 6.2, (8.5), (9.0), 10.5.

Using the data described above, with remaining observations being censored at 12 months, Table 14-3 shows the eight distinct times of death, (t_j), the number in each group at risk prior to the death, (a_j+b_j, c_j+d_j), the number of deaths at time t_j, (a_j, c_j), and the number of subjects lost to follow-up in the subsequent interval (λ_j). The entries in this table are similar to those given for the Kaplan-Meier lifetable shown in Table 14-2. Note in Table 14-3, however, that the observations from two groups have been combined with the net result being more intervals. The entries in Table 14-3 labeled a_j+b_j, c_j+d_j, a_j+c_j, and b_j+d_j become the entries in the eight 2×2 tables shown in Table 14-4.

Table 14-3
Comparison of Survival Data for a Control Group and an Intervention Group Using the Mantel-Haenszel Procedure

Rank	Event Times	Intervention			Control			Total	
j	t_j	a_j+b_j	a_j	λ_j	c_j+d_j	c_j	λ_j	a_j+c_j	b_j+d_j
1	0.5	20	0	0	20	1	1	1	39
2	1.0	20	1	0	18	0	0	1	37
3	1.5	19	0	2	18	2	1	2	35
4	3.0	17	0	1	15	1	2	1	31
5	4.5	16	1	0	12	0	0	1	27
6	4.8	15	0	1	12	1	0	1	26
7	6.2	14	0	1	11	1	2	1	24
8	10.5	13	0	13	8	1	7	1	20

a_j+b_j = number of subjects at risk in the intervention group prior to the death at time t_j
c_j+d_j = number of subjects at risk in the control group prior to the death at time t_j
a_j = number of subjects in the intervention group who died at time t_j
c_j = number of subjects in the control group who died at time t_j
λ_j = number of subjects who were lost or censored between time t_j and time t_{j+1}
a_j+c_j = number of subjects in both groups who died at time t_j
b_j+d_j = number of subjects in both groups who are alive minus the number who died at time t_j

Table 14-4
Eight 2 × 2 Tables Corresponding to the Event Times Used in the
Mantel-Haenszel Statistic in Survival Comparison of Intervention
(I) **and Control** *(C)* **Groups**

1.	(0.5 mo)*	D†	A‡	R§	5.	(4.5 mo)	D	A	R
	I	0	20	20		I	1	15	16
	C	1	19	20		C	0	12	12
		1	39	40			1	27	28
2.	(1.0 mo)	D	A	R	6.	(4.8 mo)	D	A	R
	I	1	19	20		I	0	15	15
	C	0	18	18		C	1	11	12
		1	37	38			1	26	27
3.	(1.5 mo)	D	A	R	7.	(6.2 mo)	D	A	R
	I	0	19	19		I	0	14	14
	C	2	16	18		C	1	10	11
		2	35	37			1	24	25
4.	(3.0 mo)	D	A	R	8.	(10.5 mo)	D	A	R
	I	0	17	17		I	0	13	13
	C	1	14	15		C	1	7	8
		1	31	32			1	20	21

*Number in parentheses indicates time, t_j, of a death in either group
†D = number of subjects who died at time t_j
‡A = number of subjects who are alive between time t_j and time t_{j+1}
§R = number of subjects who were at risk before death at time t_j $(R = D+A)$

The Mantel-Haenszel statistic can be computed from these eight 2 × 2 tables (Table 14-4) or directly from Table 14-3. The term $\Sigma_{j=1}^{8} a_j = 2$ since there are only two deaths in the intervention group. Evaluation of the term $\Sigma_{j=1}^{8} E(a_j) = 20/40 + 20/38 + 2\times19/37 + 17/32 + 16/28 + 15/27 + 14/25 + 13/21$ or $\Sigma_{j=1}^{8} E(a_j) = 4.89$. The value for $\Sigma_{j=1}^{8} V(a_j)$ is computed as

$$\Sigma_{j=1}^{8} V(a_j) = \frac{(1)(39)(20)(20)}{(40)^2(39)} + \frac{(1)(37)(20)(18)}{(38)^2(37)} + ...$$

This term is equal to 2.21. The computed statistic is $MH = (2 - 4.89)^2/2.21 = 3.78$. This is not significant at the 0.05 significance level for a chi-square statistic with one degree of freedom. The MH statistic can also be used when the precise time of death is unknown. If death is known to have occurred within an interval, 2 × 2 tables can be created for each interval and the method applied. For small samples, the MH statistic using a continuity correction is sometimes used. The modified numerator is

$$\{|\ \Sigma_{j=1}^{K}\ [a_j\ -\ E(a_j)]\ |\ -\ 0.5\}^2$$

where the vertical bars denote the absolute value. For the example, apply-ing the continuity correction reduces the *MH* statistics from 3.76 to 2.59.

Gehan[10] developed another procedure for comparing the survival experience of two groups of subjects by generalizing the Wilcoxon rank statistic. The Gehan statistic is based on the ranks of the observed sur-vival times. The null hypothesis, $P_I(t) = P_C(t)$, is tested. The procedure, as originally developed, involved a complicated calculation to obtain the variance of the test statistic. Mantel[14] proposed a simpler version of the variance calculation, which is most often used.

The N_I observations from the intervention group and the N_C obser-vations from the control group must be combined into a sequence of $N_C + N_I$ observations and ranked in ascending order. Each observation is compared to the remaining $N_C + N_I - 1$ observation and given a score U_i which is defined as follows:

U_i = (number of observations ranked definitely less than the ith observation) − (number of observations ranked definitely greater than the ith observation).

The survival outcome for the ith subject will certainly be larger than that for subjects who died earlier. For censored subjects, it cannot be determined whether survival time would have been less than, or greater than, the ith observation. This is true whether the ith observation is a death or a loss. Thus, the first part of the score U_i assesses how many deaths definitely preceded the ith observation. The second part of the U_i score considers whether the current, i^{th}, observation is a death or a loss. If it is a death, it definitely precedes all later ranked observations regardless of whether the observations correspond to a death or a loss. If the ith obser-vation is a loss, it cannot be determined whether the actual survival time will be less than or greater than any succeeding ranked observation, since there was no opportunity to observe the ith subject completely.

Table 14-5 ranks the 40 combined observations $(N_C = 20, N_I = 20)$ from the example used in the discussion of the Mantel-Haenszel statistic. The last 19 observations were all censored at 12 months of follow-up, seven in the control group and 12 in the intervention group. The score U_I is equal to the zero observations definitely less than 0.5 months, minus the 39 observations which are definitely greater than 0.5 months, or $U_I = -39$. The score U_2 is equal to the one observation definitely less than the loss at 0.6 months, minus none of the observations which will be definitely greater, since at 0.6 months the observation was a loss, or $U_2 = 1$. U_3 is equal to the one observation (0.5 months) definitely less than 1.0 month minus the 37 observations definitely greater than 1.0 month giving $U_3 = -36$. The last 19 observations will have scores of 9 reflecting the nine deaths which definitely precede censored observations at 12.0 months.

The Gehan statistic, G, involves the scores U_i and is defined as

$$G = W^2/V(W)$$

where $\quad W = \Sigma\, U_i \quad\quad$ (U_i's in control group only)

and $\quad\quad V(W) = \dfrac{N_C N_I}{(N_C + N_I)(N_C + N_I - 1)}\ \Sigma_{i=1}^{N_C + N_I}\ (U_i^2)$

Table 14-5
Example of Gehan Statistic Scores U_i
for Intervention (I) and Control (C) Groups

Observation i	Ranked Observed Time	Group	Definitely Less	Definitely More	U_i
1	0.5	C	0	39	−39
2	(0.6)*	C	1	0	1
3	1.0	I	1	37	−36
4	1.5	C	2	35	−33
5	1.5	C	2	35	−33
6	(1.6)	I	4	0	4
7	(2.0)	C	4	0	4
8	(2.4)	I	4	0	4
9	3.0	C	4	31	−27
10	(3.5)	C	5	0	5
11	(4.0)	C	5	0	5
12	(4.2)	I	5	0	5
13	4.5	I	5	27	−22
14	4.8	C	6	26	−20
15	(5.8)	I	7	0	7
16	6.2	C	7	24	−17
17	(7.0)	I	8	0	8
18	(8.5)	C	8	0	8
19	(9.0)	C	8	0	8
20	10.5	C	8	20	−12
21	(11.0)	I	9	0	9
22–40	(12.0)	12I,7C	9	0	9

*Parentheses indicate censored observation.

The G statistic has approximately a chi-square distribution with one degree of freedom. Therefore, the critical value is 3.84 at the 5% significance level and 6.63 at the 1% level.

In the example, $W = -87$ and the variance $V(W) = 2314.35$. Thus, $G, = (-87)^2/2314.35$ or 3.27 for which the P-value is equal to 0.071. This is compared with the P-value of 0.052 obtained using the Mantel-Haenszel statistic.

The Gehan and the Mantel-Haenszel statistics are commonly used tests in survival analysis. The Mantel-Haenszel statistic is more powerful for survival distributions of the exponential form $P_I(t) = \{P_C(t)\}^\theta$ where $\theta \neq 1$.[14] The Gehan statistic, on the other hand, is more powerful for survival distributions of the logistic form $P(t,\theta) = e^{t+\theta}/(1+e^{t+\theta})$. In actual practice, however, the distribution of the survival curve of the study population is not known. The Gehan statistic assumes the censoring pattern to be equal in the two groups. Breslow[15] considered the case in which censoring patterns are not equal and used the same statistic G with a modified variance. This modified version should be used if the censoring patterns are radically different in the two groups. However, the concepts are similar to what has been described. Peto and Peto[16] also proposed a version of a censored Wilcoxon test which is similar to the Gehan test but has some attractive features which are useful in data monitoring (Chapter 15).

Tarone and Ware[17] point out that the Mantel-Haenszel and the Gehan tests are only two of many that could be constructed. They show that when the null hypothesis is not true, the Gehan statistic gives more weight to the early survival experience, whereas the Mantel-Haenszel weights the later experience more. Tarone and Ware indicate that other possible weighting schemes could be proposed which are intermediate to these two statistics. Thus, when survival analysis is done using both the Gehan and Mantel-Haenszel statistics, it is certainly possible to obtain different results depending on where, if at all, the survival curves separate.

There has been considerable interest in asymptotic (large sample) properties of rank tests[18-20] as well as comparisons of the various methods.[21] A more general form of the Tarone-Ware class of rank tests has been developed.[22] However, the basic concepts of rank tests can still be appreciated by the methods described above.

Covariate Adjusted Analysis

Previous chapters have discussed the rationale for taking stratification into account. If differences in important covariates or prognostic variables exist at entry between the intervention and control groups, an investigator might be concerned that the analysis of the survival experience is influenced by that difference. In order to adjust for these differences in prognostic variables, the investigator could do a stratified analysis or a covariance type of survival analysis. If these differences are not important in the analysis, the adjusted analysis will give approximately the same results as the unadjusted.

Three basic techniques for stratified survival analysis are of interest. The first compares the survival experience between the study groups within each stratum, using the methods described in the previous section. By comparing the results from each stratum, the investigator can get some indication of the consistency of results across strata and the possible interaction between strata and intervention.

The second and third methods are basically adaptations of the Mantel-Haenszel and Gehan statistics and allow the results to be accumulated over the strata. The Mantel-Haenszel stratified analysis involves dividing the population into S strata and within each stratum j, forming a series of 2×2 tables for each K_j death, where K_j is the number of deaths in stratum j. The table for the ith death in the jth stratum would be as follows:

	Dead	Alive	
Intervention	a_{ij}	b_{ij}	$a_{ij} + b_{ij}$
Control	c_{ij}	d_{ij}	$c_{ij} + d_{ij}$
	$a_{ij} + c_{ij}$	$b_{ij} + d_{ij}$	n_{ij}

The entries a_{ij}, b_{ij}, c_{ij} and d_{ij} are defined as before and

$$E(a_{ij}) = (a_{ij} + c_{ij})(a_{ij} + b_{ij})/n_{ij}$$

$$V(a_{ij}) = \frac{(a_{ij} + c_{ij})(b_{ij} + d_{ij})(a_{ij} + b_{ij})(c_{ij} + d_{ij})}{n_{ij}^2 (n_{ij} - 1)}$$

The Mantel-Haenszel statistic is

$$MH = \{\Sigma_{j=1}^{S} \Sigma_{i=1}^{K_j} a_{ij} - E(a_{ij})\}^2 / \Sigma_{j=1}^{S} \Sigma_{i=1}^{K_j} V(a_{ij})$$

which has a chi-square distribution with one degree of freedom. Analogous to the Mantel-Haenszel statistic for stratified analysis, one could compute a Gehan statistic W_j and $V(W_j)$ within each stratum. Then an overall stratified Gehan statistic is computed as

$$G = \{\Sigma_{j=1}^{S} W_j\}^2 / \Sigma_{j=1}^{S} V(W_j)$$

which also has chi-square statistic with one degree of freedom.

If there are many covariates, each with several levels, the number of strata can quickly become large, with few subjects in each. Moreover, if a covariate is continuous, it must be divided into intervals and each interval assigned a score or rank before it can be used in a stratified analysis. Cox[23] proposed a regression model which allows for analysis of censored survival data adjusting for continuous as well as discrete covariates, thus avoiding these two problems. For the special case where group assignment is the only covariate, the Cox model is essentially equivalent to the Mantel-Haenszel statistic.

One way to understand the Cox regression model is to consider a simpler parametric model. If one expresses the probability of survival to time t, denoted $S(t)$, as an exponential model, then $S(t) = e^{-\lambda t}$ where the parameter, λ, is called the force of mortality or the hazard rate. The larger the value of λ, the faster the survival curve decreases. Models have been proposed which attempt to incorporate the hazard rate as a linear function of several covariates,[24,25] eg, $\lambda(X_1, X_2, \dots, X_p) = b_1 X_1 + b_2 X_2 + \dots + b_p X_p$. Cox suggested that the hazard rate could be modeled as a function of both time and covariates, denoted $\lambda(t, X_1, X_2, \dots, X_p)$. Moreover, this hazard rate could be represented as the product of two terms, the first representing an unadjusted force of mortality $\lambda(t)$ and the second the adjustment for the linear combination of a particular covariate profile. More specifically, the Cox proportional hazard model assumes that

$$\lambda(t, X_1, X_2, \dots, X_p) = \lambda(t) \exp(b_1 X_1 + b_2 X_2 + \dots + b_p X_p)$$

One of the covariates, say X_1, might represent the intervention and the others, for example, might represent age, sex, performance status, or prior medical history. The coefficient, b_1, then would indicate whether intervention is a significant prognostic factor, ie, is effective after having adjusted for the other factors. The estimation of the regression coefficients $b_1, b_2, \dots, b_p$ is complex and goes beyond the scope of this text. However, computer programs exist in many statistical computing packages. Despite the complexity of the parameter estimation, this method is widely applied and has been studied extensively.[26-34] A recent paper by Pocock, Gore, and Kerr[35] demonstrates the value of some of these methods with cancer data.

The techniques described in this chapter as well as the extensions or generalizations referenced are powerful tools in the analysis of survival data. Perhaps none is exactly correct for any given set of data but experience indicates they are fairly robust and quite useful.

210

REFERENCES

1. Brown, B.W., and Hollander, M. *Statistics: A Biomedical Introduction.* New York: John Wiley and Sons, 1977.
2. Armitage, P. *Statistical Methods in Medical Research.* New York: John Wiley and Sons, 1977.
3. Kalbfleisch, J.D., and Prentice, R.L. *The Statistical Analysis of Failure Time Data.* New York: John Wiley and Sons, 1980.
4. Miller, R.G., Jr. *Survival Analysis.* New York: John Wiley and Sons, 1981.
5. Cox, D.R., and Oakes, D. *The Analysis of Survival Data.* New York: Chapman and Hall, 1984.
6. Kaplan, E., and Meier, P. Nonparametric estimation from incomplete observations. *J Am Stat Assoc.* 53:457–481, 1958.
7. Cutler, S., and Ederer, F. Maximum utilization of the lifetable method in analyzing survival. *J Chronic Dis.* 8:699–712, 1958.
8. Greenwood, M. The natural duration of cancer. Reports on Public Health and Medical Subjects, No. 33, 1926, His Majesty's Stationery Office.
9. Thomas, D.G., Breslow, N., and Gart, J. Trend and homogeneity analysis of proportions and life table data. *Computers Biomed Res.* 10:373–381, 1977.
10. Gehan, E. A generalized Wilcoxon test for comparing arbitrarily single censored samples. *Biometrika* 52:203–223, 1965.
11. Mantel, N. Evaluation of survival data and two new rank order statistics arising in its consideration. *Cancer Chemother Rep.* 50:163–170, 1966.
12. Cochran, W. Some methods for strengthening the common χ^2 tests. *Biometrics* 10:417–451, 1954.
13. Mantel, N., and Haenszel, W. Statistical aspects of the analysis of data from retrospective studies of disease. *J Natl Cancer Inst.* 22:719–748, 1959.
14. Mantel, N. Ranking procedures for arbitrarily restricted observations. *Biometrics* 23:65–78, 1967.
15. Breslow, N. A generalized Kruskal-Wallis test for comparing K samples subject to unequal patterns of censorship. *Biometrika* 57:579–594, 1970.
16. Peto, R., and Peto, J. Asymptotically efficient rank invariant test procedures. *J R Stat Soc.* Series A 135:185–207, 1972.
17. Tarone, R., and Ware, J. On distribution-free tests for equality of survival distributions. *Biometrika* 64:156–160, 1977.
18. Oakes, D. The asymptotic information in censored survival data. *Biometrika* 64:441–448, 1977.
19. Prentice, R.L. Linear rank tests with right censored data. *Biometrika* 65:167–179, 1978.
20. Schoenfeld, D. The asymptotic properties of non-parametric tests for comparing survival distributions. *Biometrika* 68:316–319, 1981.
21. Leurgans, S.L. Three classes of censored data rank tests: strengths and weakness under censoring. *Biometrika* 70:651–658, 1983.
22. Harrington, D.P., and Fleming, T.R. A class of rank test procedures for censored survival data. *Biometrika* 69:553–566, 1982.
23. Cox, D.R. Regression models and lifetables. *J R Stat Soc.* Series B 34:187–202, 1972.
24. Zelen, M. Application of exponential models to problems in cancer research. *J R Stat Soc.* Series A 129:368–398, 1966.

25. Prentice, R.L., and Kalbfleisch, J.D. Hazard rate models with covariates. *Biometrics* 35:25–39, 1979.

26. Kalbfleisch, J.D., and Prentice, R.L. Marginal likelihoods based on Cox's regression and life model. *Biometrika* 60:267–278, 1973.

27. Breslow, N. Covariance analysis of censored survival data. *Biometrics* 30:89–99, 1974.

28. Breslow, N. Analysis of survival data under the proportional hazards model. *Internatl Stat Rev.* 43:45–58, 1975.

29. Kay, R. Proportional hazard regression models and the analysis of censored survival data. *J R Stat Soc.* Series C 26:227–237, 1977.

30. Prentice, R.L., and Gloeckler, L.A. Regression analysis of grouped survival data with application to breast cancer. *Biometrics* 34:57–67, 1978.

31. Efron, B. The efficiency of Cox's likelihood function for censored data. *J Am Stat Assoc.* 72:557–565, 1977.

32. Tsiatis, A.A. A large sample study of Cox's regression model. *Annals Stat.* 9:93–108, 1981.

33. Schoenfeld, D. Chi-squared goodness-of-fit tests for the proportional hazards regression model. *Biometrika* 67:145–153, 1980.

34. Storer, B.E., and Crowley, J. Diagnostics for Cox regression and general conditional likelihoods. *J Am Stat Assoc.* 80:139–147, 1985.

35. Pocock, S.J., Gore, S.M., and Kerr, G.R. Long term survival analysis: the curability of breast cancer. *Stat Med.* 1:93–104, 1982.

CHAPTER 15

Monitoring Response Variables

The investigator's ethical responsibility to the study subjects demands that results be monitored during the trial. If data partway through the trial indicate that the intervention is harmful to the subjects, early termination of the trial should be considered. If these data demonstrate a clear benefit from the intervention, the trial may also be stopped early because to continue would be unethical. In addition, if differences in primary—and possibly secondary—response variables are so unimpressive that the prospect of a clear result is extremely unlikely, it may not be justifiable in terms of time, money and effort to continue the trial. Finally, monitoring of response variables can identify the need to collect additional data in order to clarify questions of benefit or toxicity that may arise during the trial. In order to fulfill the monitoring function, the data must be collected and processed in a timely fashion as the trial progresses. Data monitoring would be of limited value if conducted only at a time when all or most of the data had been collected. The specific issues related to monitoring of recruitment, compliance, and quality control are covered in other chapters and will not be discussed here.

FUNDAMENTAL POINT

During the trial, response variables need to be monitored for early dramatic benefits or potential harmful effects. Preferably, monitoring should be done by a person or group independent of the investigator. Although many techniques are available to assist in monitoring, none of them should be used as the sole basis in the decision to stop or continue the trial.

DATA MONITORING COMMITTEE

Data monitoring is not simply a matter of looking at tables or results of statistical analysis of the primary outcome but is an active process in

213

which additional tabulations and analysis are suggested and developed as a result of ongoing review. Data monitoring also involves an interaction between the individuals responsible for collating, tabulating, and analyzing the data. For single center studies, the monitoring responsibility could, in principle, be assumed by the investigator. However, he may find himself in a difficult situation. While monitoring the data, he may discover that the results trend in one direction or the other while subjects are still being enrolled. Presumably, he recruits subjects to enter a trial on the basis that he favors neither intervention nor control. Knowing that a trend exists may make it difficult for him to continue enrolling subjects. It is also difficult for the investigator to follow, evaluate, and care for the subjects in an unbiased manner knowing that a trend exists. Furthermore, the credibility of the trial is enhanced if an independent person monitors the response variable data. Because of these considerations we, as well as others,[1-3] recommend that the individuals who monitor a clinical trial have no formal involvement with the subjects or the investigators.

Except for small, short-term studies, when one or two knowledgeable individuals may suffice, the responsibility for monitoring response variable data is usually placed with an independent group with expertise in various disciplines. The independence protects the members of the monitoring committee from being influenced in the decision-making process by investigators or study subjects. The committee would usually include experts in the relevant clinical fields or specialties, individuals with experience in the conduct of clinical trials, epidemiologists, and biostatisticians knowledgeable in design and analysis.

No simple formula can be given for how often a monitoring committee should meet. The frequency may vary depending on the phase of the trial. Subject recruitment, follow-up and closeout phases require different levels of activity. Meetings should not be so frequent that little new data are accumulated in the interim, given the time and expense of convening a committee. If potential toxicity of one of the interventions becomes an issue during the trial, special meetings may be needed. In many long-term clinical trials, the monitoring committees have met regularly at four- to six-month intervals, with additional meetings as needed. Between committee meetings, the person or persons responsible for collating, tabulating, and analyzing the data assume the responsibility for monitoring unusual situations which need to be brought to the attention of the monitoring committee.

REPEATED TESTING FOR SIGNIFICANCE

In the discussion on sample size (Chapter 7) the issue of testing several hypotheses was raised and referred to as the "multiple testing" problem.

Similarly, while repeated significance testing of accumulating data is essential to the monitoring function, it does have statistical implications which have been described by several authors.[4-9] If the null hypothesis, H_0, of no difference between two groups is, in fact, true, and repeated tests of that hypothesis are made at the same level of significance using accumulating data, the probability that, at some time, the test will be called significant by chance alone will be larger than the significance level selected. That is, the rate of incorrectly rejecting the null hypothesis will be larger than what is normally considered to be acceptable.

In a clinical trial in which the subject response is known relatively soon after entry, the difference in rates between two groups may be compared repeatedly as more subjects are added to the study. The usual test statistic for comparing two proportions used is the chi-square test or the equivalent normal test statistic.[7] The null hypothesis is that the true response rates or proportions are equal. If a significance level of 5% is selected and the null hypothesis, H_0, is tested only once, the probability of rejecting H_0 if it is true is 5% by definition. However, if H_0 is tested twice, first when one-half of the data are known and then when all the data are available, the probability of incorrectly rejecting H_0 is increased from 5% to 8%.[6,7] If the hypothesis is tested five times, with one-fifth of the subjects added between tests, the probability of finding a significant result if the usual statistic for the 5% significance level is used becomes 14%. For 10 tests, this probability is almost 20%. In a clinical trial in which long-term survival experience is the primary outcome, repeated tests might be done as more information becomes known about the enrolled subjects. Canner[3] performed computer simulations of such a clinical trial in which both the control group and intervention group event rates were assumed to be 30% at the end of the study. He performed 2000 replications of this simulated experiment. He found that if 20 tests of significance are done within a trial, the chance of crossing the 5% significance level boundaries ($Z = \pm 1.96$) is, on the average, 35%. Thus, in either of the situations described, repeated testing of accumulating data without taking into account the number of tests increases the overall probability of incorrectly rejecting H_0 to levels which could be unacceptable. If the repeated testing continues indefinitely, the null hypothesis is certain to be rejected eventually. Although it is unlikely that a large number of repeated tests will be done, even five or ten can lead to a misinterpretation of the results of a trial when the multiple testing issue is ignored.

In general, the solution to multiple testing is to adjust the α used in a single test so that the overall significance level for the trial remains at the desired level. It has been suggested that a trial should not be terminated early unless the difference is very significant.[8] More formal monitoring techniques are reviewed later in this chapter. They include the "classical sequential methods," group sequential methods and stochastic curtailed sampling procedures.

DECISION FOR EARLY TERMINATION

There are three major reasons for terminating a trial earlier than scheduled.[9] The trial may show serious adverse effects in the entire intervention group or in a subgroup. In addition, the trial may indicate greater-than-expected beneficial effects. Finally, it may become clear that a statistically significant difference by the end of the study is improbable. For a variety of reasons, a decision to terminate a study early must be made with a great deal of caution and in the context of all pertinent data. A number of issues or factors must be considered thoroughly as part of the decision process.[10-14]

1. Possible differences in prognostic factors between the two groups at baseline should be explored and necessary adjustments made in the analysis.
2. Any chance of bias in the assessment of response variables must be taken into account, especially when the trial is not double-blind.
3. The possible impact of missing data should be evaluated. For example, could the conclusions be reversed if the experience of subjects with missing data from one group were different from the experience of subjects with missing data from the other group?
4. Differential concomitant intervention and levels of subject compliance should be evaluated for their possible impact.
5. Potential side effects and outcomes of secondary response variables should be considered in addition to the outcome of the primary response variable.
6. Internal consistency should be examined. Are the results consistent across subgroups and the various outcome measures? In a multicenter trial, the monitoring committee should assess whether the results are consistent across centers. Before stopping too soon, the committee should make certain that the outcome is not due to unusual experience in only one or two centers.
7. In long-term trials, the experience of the study groups over time should be explored. Survival analysis techniques (Chapter 14) partly address this issue.
8. The outcomes of similar trials should be reviewed.
9. The impact of early termination on the credibility of the results is an important factor.

The early termination of a clinical trial can be difficult,[15-18] not only because the issues involved may be complex and the study complicated, but because the final decision often lies with the consensus of a committee. The statistical methods discussed in this chapter are useful guides in this process but should not be viewed as absolute rules. Only limited information is available in the published literature on the data-monitoring experience of clinical trials.[19]

One of the earlier clinical trials conducted in the United States illustrates how controversial the decision for early termination may be. The University Group Diabetes Program (UGDP) was a placebo-control, randomized, double-blind trial designed to test the effectiveness of four interventions used in the treatment of diabetes. The primary measure of efficacy was the degree of retinal damage. The four interventions were: a fixed dose of insulin, a variable dose of insulin, tolbutamide and phenformin. The tolbutamide group was stopped early because the monitoring committee felt the drug could be harmful and did not appear to have any benefit.[20] An excess in cardiovascular mortality was observed in the tolbutamide group as compared to the placebo group (12.7% vs. 4.9%) and the total mortality was in the same direction (14.7% vs. 10.2%). Analysis of the distribution of the baseline factors known to be associated with cardiovascular mortality revealed an imbalance, with subjects in the tolbutamide group being at higher risk. This, plus questions about the assignment of cause of death, drew considerable criticism. Later, the phenformin group was also stopped because of excess mortality in the control group (15.2% vs. 9.4%).[21]

The controversy led to a review of the data by an independent group of statisticians. Although they basically concurred with the decisions made by the UGDP monitoring committee,[22] the debate over the study and its conclusion continues.[23]

The decision-making process during the course of the Coronary Drug Project,[24] a long-term randomized, double-blind, multicenter study which compared the effect on total mortality of several lipid-lowering drugs (high- and low-dose estrogen, dextrothyroxine, clofibrate, nicotinic acid) against placebo has been reviewed.[3,11,25-28] Three of the drug interventions were terminated early because of potential side effects and no apparent benefit. One of the issues in the discontinuation of the high dose estrogen and dextrothyroxine interventions[25,26] concerned subgroups of subjects. In some, the interventions appeared to cause increased mortality, in addition to having a number of adverse effects. In others, the adverse effects were present, but mortality was only slightly reduced or unchanged. After considerable debate, both treatments were discontinued. The adverse effects were felt to more than outweigh the minimal benefit in selected subgroups. Also, positive subgroup trends in the dextrothyroxine arm were not maintained over time. The low dose estrogen intervention[27] was discontinued because there was the question of major toxicity. Furthermore, it was extremely improbable that a significant difference could have been obtained had the study continued to its scheduled termination. Using the data available at the time, the number of future deaths in the control group was projected. This indicated that there had to be almost no further deaths in the intervention group for a significance level of 5% to be reached.

The Coronary Drug Project experience also warns against the dangers of stopping too soon.[11,28] In the early months of the study, clofibrate appeared to be beneficial, with the significance level exceeding 5% on three occasions (Figure 15-1). However, because of the repeated testing issue, the decision was made to continue the study and closely monitor the results. The early difference was not maintained, and at the end of the trial the drug showed no benefit over placebo. It is notable that the mortality curves shown in Figure 15-2 do not suggest the wide swings observed in the interim analyses shown in Figure 15-1. The fact that patients were entered over a period of time and thus had various lengths of follow-up at any given interim analysis, explains the difference between the two types of analyses. (See Chapter 14 for a discussion of survival analysis.)

The Nocturnal Oxygen Therapy Trial was a randomized, multicenter clinical trial comparing two levels of oxygen therapy in subjects with

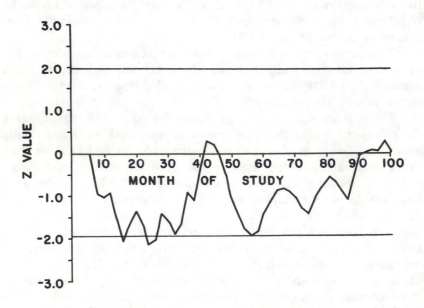

Figure 15-1 Interim survival analyses comparing mortality in clofibrate and placebo treated subjects in the Coronary Drug Project. A positive Z value favors placebo.[11] Reprinted by permission of Elsevier Science Publishing Co., Inc.

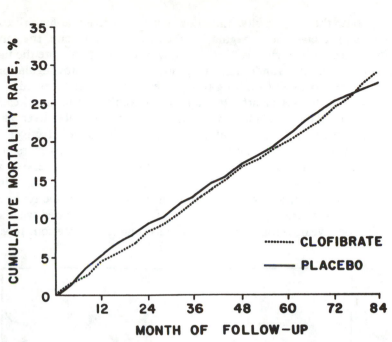

Figure 15-2 Cumulative mortality curves comparing clofibrate and placebo-treated subjects in the Coronary Drug Project.[11] Reprinted by permission of Elsevier Science Publishing Co., Inc.

advanced chronic obstructive lung disease.[29] While mortality was not considered as the primary outcome in the design, a strong mortality difference emerged during the trial, notably in one particular subgroup. Before any decision was made, the participating clinical centers were surveyed to ensure that the mortality data were as current as possible. A delay in reporting mortality was discovered. When all the deaths were considered, the trend disappeared. The earlier results were an artifact caused by incomplete mortality data. Although a significant mortality difference ultimately emerged, the results were similar across subgroups.

A scientific and ethical issue was raised in the Diabetic Retinopathy Study, a randomized trial of 1758 subjects with proliferative retinopathy.[30] Each subject had one eye randomized to photocoagulation and the other to standard care. After two years of a planned five year follow-up, a highly significant difference in the incidence of blindness was observed (16.3% vs. 6.4%) in favor of photocoagulation.[31] Since the long-term efficacy of this new therapy was not known, the early benefit could possibly have been negated by subsequent adverse reactions. After much debate, the data-monitoring committee decided to continue the

220

trial, publish the early results, and allow any untreated eye at high risk of blindness to receive photocoagulation therapy.[32-33] In the end, the early treatment benefit was sustained over a longer follow-up, despite the fact that some of the eyes randomized to control received photocoagulation. Furthermore, no significant long-term adverse effect was observed.

The Beta-Blocker Heart Attack Trial was another example of early termination.[34] This randomized placebo control trial enrolled over 3800 subjects with a recent myocardial infarction to evaluate the effectiveness of propranolol in reducing mortality. After an average of two years of a planned three year follow-up, a mortality difference was observed, as shown in Figure 15-3. The results were statistically significant, allowing for repeated testing, and would, with high probability, not be reversed during the next year.[13] The data-monitoring committee debated whether the additional year of follow-up would add valuable information. It was

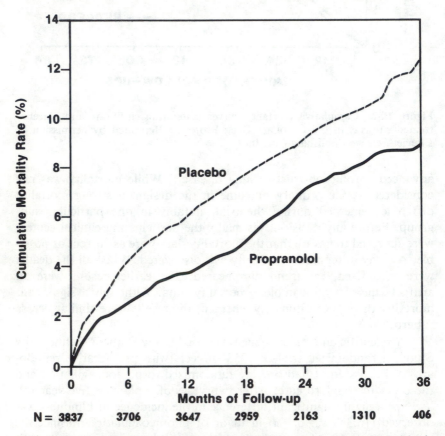

Figure 15-3 Cumulative mortality curves comparing propranolol and placebo in the Beta-Blocker Heart Attack Trial. Reproduced with permission of Beta-Blocker Heart Attack Study Group.[34]

argued that there would be too few events in the last year of the trial to provide a good estimate of the effect of propranolol treatment in the third and fourth year of therapy. Thus, the committee decided that prompt publication of the observed benefit was more important than waiting for the marginal information yet to be obtained.

Trials may continue to their scheduled termination even though interim results are very persuasive[35] or so similar that almost surely no significant results will emerge.[36-38] In one case, early significant results did not override the need for getting long-term experience with an intensive intervention strategy.[35] Another trial[37] involved a life-style modification in altering smoking and cholesterol levels and drug therapy for altering blood pressure. Although early results showed no trends, it was also not clear how long intervention needed to be continued before the applied risk factor modifications would take effect. It was argued that late favorable results could still emerge. In another trial, the medical care group had such a favorable survival experience that there was little room for improvement by immediate surgical intervention.[38]

In all of these studies, the decisions were difficult and involved many analyses, thorough review of the literature, and an understanding of the biological processes. As described above, a number of questions must be answered before serious consideration should be given to early termination.

DECISION TO EXTEND A TRIAL

The issue about extending a trial beyond the original sample size or planned period of follow-up may arise.[16] Suppose the mortality rate over a two-year period in the control group is assumed to be 40%. (This estimate may be based on data from another study involving a similar population.) Also specified is that the sample size should be large enough to detect a 25% reduction due to the intervention, with a two-sided significance level of 5% and a power of 90%. The total sample size is, therefore, approximately 960. However, say that early in the study, the mortality rate in the control group appears somewhat lower than anticipated; closer to 30%. This difference may result from a change in the study population, selection factors in the trial, or new concomitant therapies. If no design changes are made, the intervention would have to be more effective (30% reduction rather than 25%) for the difference between groups to be detected with the same power. Alternatively, the investigator would have to be satisfied with approximately 75% power of detecting the originally anticipated 25% reduction in mortality. If it is unreasonable to expect a 30% benefit and if a 75% power is unacceptable, the design needs modification. Given the lower control group mortality rate, approximately 1450 subjects would be required to detect a

25% reduction in mortality due to intervention, while maintaining a power of 90%. Another possible option is to extend the length of follow-up, which would increase the overall event rate. A combination of these two approaches can also be tried.

In a trial of antenatal steroid administration,[39] the incidence of infant respiratory distress in the control group was much less than anticipated. Early in the study, the investigators decided to increase the sample size by extending the recruitment phase. In another trial, the protocol specifically called for increasing the sample size if the control group event rate was less than assumed.[40]

In the above situations, only data from the control group were used. No knowledge of what was happening in the intervention group was needed. However, if the intervention group results are not used in the recalculations, then an increase in sample size could be recommended when the observed difference between the intervention and control groups is actually larger than originally expected. Thus, in the hypothetical example, if early data really did show a 30% benefit from intervention, an increased sample size might not be needed to maintain the desired power of 90%. For this reason, one would not like to make a recommendation about extension without also considering the observed effect of intervention. Computing conditional power is one way of incorporating these results. Conditional power is the probability that the test statistic will be larger than the critical value, given that a portion of the statistic is already known from the observed data. As in other power calculations, the assumed true difference in response variables between groups must be specified. When the early intervention experience is better than expected, the conditional power will be large. When the intervention is doing worse than anticipated, the conditional power will be small unless the sample size is increased. The conditional power concept utilizes knowledge of outcome in both the intervention and control groups and is, therefore, controversial. Nevertheless, the concept attempts to quantify the decision to extend.

Towards the scheduled end of a study, the investigator may find that he has nearly statistically significant results. He may be tempted to extend or expand the trial in an effort to make the test statistic significant. Such a practice is not recommended. A strategy of extending assumes that the observed relative differences in rates of response will continue. The observed differences which are projected for a larger sample may not hold. In addition, because of the multiple testing issue and the design change, the significance level should be adjusted downward. However, appropriate adjustments in the significance level to account for the design changes may not easily be determined. Since a more extreme significance level should be employed, and since future responses are uncertain, extension may leave the investigator without the expected benefits.

Whatever adjustments are made to either sample size or the length of follow-up should be done as early in the trial as possible. Early adjustments would diminish the criticism that the monitoring committee waited until the last minute to see whether the results would achieve some prespecified significance level before changing the study design.

STATISTICAL METHODS USED IN MONITORING

In this section, some statistical methods currently available for monitoring the accumulating data in a clinical trial will be reviewed. The methods address whether the trial should be terminated early or continued to its planned termination. No single statistical test or monitoring procedure ought to be used as a strict rule for decision-making, but rather as one piece of evidence to be integrated with other evidence. Most methods are very specific in their applications. Therefore, it is difficult to make a single recommendation about which should be used. However, the following methods, when applied appropriately, can be useful guides in the decision-making process.

Classical sequential methods, a modification generally referred to as group sequential methods, and curtailed testing procedures are discussed below. Other approaches are also briefly considered. These methods are given more mathematical attention in recent review articles.[41-43]

Classical Sequential Methods

The aim of the classical sequential design is to minimize the number of subjects that must be entered into a study. The decision to continue to enroll subjects depends on results from those already entered. Most of the sequential methods assume that the response variable outcome is known in a short time relative to the duration of the trial. Therefore, for many trials involving acute illness, these methods are applicable. For studies involving chronic diseases, classical sequential methods have not been as useful. However, recent developments have made it possible to apply these methods in the analyses of survival data. Although the sequential approaches have design implications, we have delayed discussing any details until this chapter because they really focus on monitoring accumulating data. Even if, during the design of the trial, consideration were not given to sequential methods, they could still be used to assist in the data monitoring or the decision-making process. Detailed discussions of classical sequential methods are given by Armitage[44] and Whitehead.[45]

The sequential analysis method as originally developed by Wald[46] and applied by Armitage[44] to the clinical trial involves repeated testing of data in a single experiment. The method assumes that the only decision to be made is whether the trial should continue or be terminated because one of the groups is responding significantly better, or worse, than the other. This classical sequential decision rule is called an "open plan" by Armitage because there is no guarantee of when a decision to terminate will be reached. Strict adherence to the "open plan" would mean that the study could not have a fixed sample size. Very few clinical trials use the "open" or classical sequential design because there is no certainty of ever reaching a point at which the trial would be stopped. The method also requires data to be paired, one observation from each group. In many instances, the pairing of subjects is not appealing because the paired subjects may be very different and may not be "well matched" in important prognostic variables. If stratification is attempted in order to obtain better matched pairs, each stratum with an odd number of subjects would have one unpaired subject. Furthermore, the requirement to monitor the data after every pair may be impossible or unnecessary for many clinical trials.

An example of an "open" sequential rule for binary response variables is shown in Figure 15-4 for a two-sided test of a hypothesis with $\alpha = 0.05$ and power = 0.95. The horizontal axis represents the number of paired observations and the vertical axis the excess number of times the intervention subject is superior to the paired control subject. The excess number of times the intervention is superior is the number of pairs in which the intervention is better minus the number of pairs in which the control is better. As seen in Figure 15-4, this net intervention effect may be positive or negative. The null hypothesis states that if the intervention is not different from the control, it will be superior to the control 50% of the time. In this example, the two-sided alternative hypothesis is that the probability of the intervention's being superior is 0.6 and the probability that intervention is inferior is 0.6. The resulting plot of this decision rule looks like a "pant leg"; staying inside the "pant leg" means continuation of the study. Hitting the inner boundary would mean stopping the study and "accepting," or more appropriately, failing to reject the null hypothesis, H_0. Hitting the upper boundary would mean rejecting H_0, thereby implying that the intervention is superior to the control. Hitting the lower boundary would mean that the control is superior to the intervention. For a one-sided hypothesis, there would be only one pant leg somewhat wider than either of the pant legs in Figure 15-4.

Silverman and colleagues[47] used an "open plan" in a trial of the effects of humidity on survival in infants with low birth weight. At the end of 36 months, 181 pairs of infants had been enrolled; 52 of the pairs had a discrepant outcome. Nine infants were excluded because they were unmatched and 16 pairs were excluded because of a mismatch. The study had to be terminated without a clear decision because it was no longer feasible to continue the trial. This study illustrates the difficulties inherent in the classical sequential design.

Armitage[48] introduced the restricted or "closed" sequential design to assure that a maximum limit is imposed on the number of subjects $(2N)$ to be enrolled. As with the "open plan," the data must be paired using one observation from each study group. The boundaries are determined so that the design has specified levels of α and $1-\beta$. The boundaries of the restricted sequential plan for a dichotomous response variable are illustrated in Figure 15-5 for the same two-sided hypothesis as used in Figure 15-4. The horizontal axis represents the number of paired observations and the vertical axis represents the excess number of times that

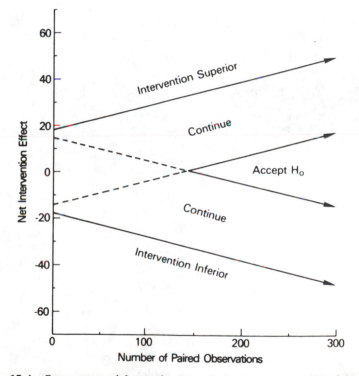

Figure 15-4 Open sequential stopping boundaries for accumulating data where the null hypothesis (H_0) is no difference between two groups and the alternative is the probability of the intervention being superior (0.6) or being inferior (0.6), with significance level of 0.05 and power of 0.95. Net intervention effect is the number of pairs intervention is superior minus the number of pairs intervention is inferior.*

---

*After Armitage.[44]

subjects in the intervention group do better than the paired control subjects or the net intervention effect. The boundary for "acceptance" of H_0, which has changed from Figure 15-4, is now represented by the vertical line for N maximum "pairs" of subjects. This boundary may also be represented by the wedge (as shown) with lines of slope one, because, whenever the plot of the observed data gets inside the wedge, it must eventually hit the vertical boundary for "acceptance" of H_0. For a specific alternative hypothesis, Armitage gives the appropriate boundary parameters for the binomially and normally distributed response variables and the maximum number of "pairs" of subjects needed. The restricted plan was used in a comparison of two interventions in subjects with ulcerative colitis.[49] In that trial, the upper boundary was crossed, demonstrating short-term clinical benefit of corticosteroids over sulphasalazine therapy.

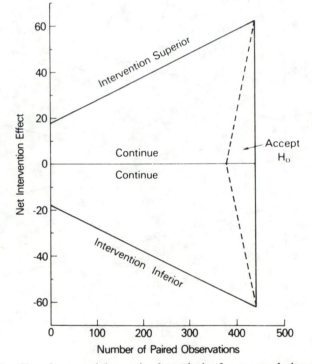

Figure 15-5 Closed sequential stopping boundaries for accumulating data where the null hypothesis *(H_0)* is no difference between two groups and the alternative is the probability of the intervention being superior (0.6) or being inferior (0.6), with significance level of 0.05 and power of 0.95. Net intervention effect is the number of pairs intervention is superior minus the number of pairs intervention is inferior.*

---

*After Armitage.[44]

Another solution to the repeated testing problem, called "repeated significance tests," was proposed by McPherson and Armitage[50] and is also described by Armitage.[44] Although different theoretical assumptions are used, this approach has features similar to the restricted sequential model. That is, the observed data must be paired, and the maximum number of pairs to be considered can be fixed. While the method for determining the boundary parameters differs, the form of the repeated significance boundary is similar to that shown in Figure 15-5 for the restricted plan. Triangular boundaries,[51,52] as well as other modifications to the Armitage restricted plan,[53] have also been proposed.

The methods described above can in some circumstances be applied to interim analyses of censored survival data.[54-64] If subjects simultaneously enter a clinical trial and there is no loss to follow-up, information on interim analyses is said to be "progressively censored." Sequential methods for this situation have been developed using, for example, modified rank statistics. In fact, most subjects are not entered into a trial simultaneously but in a staggered fashion. The log rank statistic may also be used in this situation.[60]

The classical sequential approach has not been widely used, even in clinical trials where the time to the event is known almost immediately. One major reason perhaps is the requirement of analysis after every pair of outcomes or events. For many clinical trials, this is not necessary or even feasible if the data are monitored by a committee which has regularly scheduled meetings. In addition, classical sequential boundaries require an alternative hypothesis to be specified, a feature not demanded by conventional statistical tests for the rejection of the null hypothesis.

Group Sequential Methods

Because of dissatisfaction with classical sequential methods, other approaches to the repeated testing problem have been proposed. Ad hoc rules have been suggested which attempt to ensure a conservative interpretation of interim results. One method is to use a critical value of 2.6 at each interim look as well as in the final analyses.[37] Another approach,[8,65] referred to here as the Haybittle-Peto procedure, favors using a large critical value, such as $Z_i = \pm 3.0$, for all interim tests ($i < N$). Then any adjustment for repeated testing at the final test ($i = N$) is negligible and the conventional critical value can be used. These methods are "ad hoc" in the sense that no precise Type I error level is guaranteed. They might, however, be viewed as precursors of the more formal procedures to be described below.

Pocock[66] modified the concept of McPherson and Armitage[50] and developed a group sequential method for clinical trials which avoids many of the limitations of classical methods. He discusses two cases of special interest; one for comparing two proportions and another for comparing mean levels of response. Pocock's method divides the subjects into a series of N equal-sized groups with $2n$ subjects in each, n assigned to intervention and n to control. N is the number of times the data will be monitored during the course of the trial. The test statistic used to compare control and intervention is computed as soon as data for the first group of $2n$ subjects are available, and recomputed when data from each successive group become known. Under the null hypothesis, the distribution of the test statistic, Z_i, is assumed to be approximately normal with zero mean and unit variance, where i indicates the number of groups ($i \leqslant N$) which have completed data. This statistic Z_i is compared to the stopping boundaries, $\pm Z'_N$, where Z'_N has been determined so that for up to N repeated tests, the overall significance level for the trial will be α. For example, if $N = 5$ and $\alpha = 0.05$ (two-sided), $Z'_N = 2.413$. This critical value is larger than the critical value of 1.96 used in a single test of hypothesis with $\alpha = 0.05$. If the statistic Z_i falls outside the boundaries on the "i"-th repeated test, the trial should be terminated, rejecting the null hypothesis. If the statistic falls inside the boundaries, the trial should be continued until $i = N$ (the maximum number of tests). When $i = N$, the trial would stop and the investigator would "accept" H_0.

O'Brien and Fleming[67] also discuss a group sequential procedure. Using the above notation, their stopping rule compares the statistic Z_i with $Z^* \sqrt{N/i}$ where Z^* is determined so as to achieve the desired significance level. For example, if $N = 5$ and $\alpha = 0.05$, $Z^* = 2.04$. If $N \leqslant 5$, Z^* may be approximated by the usual critical values for the normal distribution. One attractive feature is that the critical value used at the last test $(i = N)$ is approximately the same as that used if a single test were done.

In Figure 15-6, boundaries for the three methods described are given for $N = 5$ and $\alpha = 0.05$. If for $i < 5$ the test statistic falls outside the boundaries, the trial is terminated and the null hypothesis rejected. Otherwise, the trial is continued until $i = 5$, at which time the null hypothesis is either rejected or "accepted." The three boundaries have different early stopping properties. The O'Brien-Fleming model is unlikely to lead to stopping in the early stages. Later on, however, this procedure leads to a greater chance of stopping prior to the end of the study than the other two. Both the Haybittle-Peto and the O'Brien-Fleming boundaries avoid the awkward situation of accepting the null hypothesis when the observed statistic at the end of the trial is much larger than the conventional critical value (ie, 1.96 for a two-sided 5% significance level). If the observed statistic in Figure 15-6 is 2.3 when $i = 5$, the result would not be significant using the Pocock boundary. The large critical values used at the first few analyses for the O'Brien-Fleming boundary can be adjusted to some less extreme values (eg, 3.5) without noticeably changing the critical values used later on, including the final value.

Many data monitoring committees wish to be somewhat conservative in their interpretation of early results because of the uncertainties discussed earlier and because a few additional events can alter the results dramatically. Yet, most would like to use conventional critical values in the final analyses. With that in mind, the O'Brien-Fleming model has considerable appeal, perhaps with the adjusted boundary as described.

The group sequential methods have an advantage over the classical methods in that the data do not have to be continuously tested and individual subjects do not have to be "paired." This concept suits the data review activity of most large clinical trials where monitoring committees meet periodically. Furthermore, in many trials constant consideration of

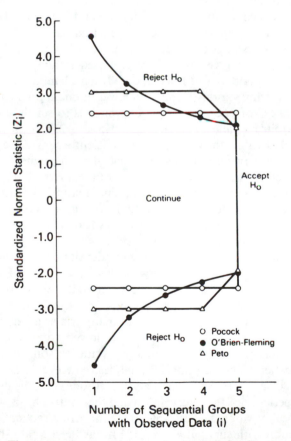

Figure 15-6 Three group sequential stopping boundaries for the standardized normal statistic (Z_i) for up to five sequential groups with two-sided significance level of 0.05.

early stopping is unnecessary. Pocock[66,68] discusses the benefits of the group sequential approach in more detail.

Further refinements to the group sequential procedures should be noted. Significance levels or "P-values" can also be computed which are adjusted for the particular boundary actually used.[69] Some attempts have also been made to incorporate into group sequential methods, physician opinion on how large a difference between intervention strategies is clinically important or a range of "therapeutic equivalence."[70] Increasing or decreasing the range of equivalence reflects the clinician's degree of preference between interventions. Additional methods to obtain confidence intervals for the results using the group sequential procedure have also been developed.[71,72]

In most trials, the main purpose is to test whether the intervention is superior to the control. It is rarely ethical to continue a study in order to prove, at the usual levels of significance, that the intervention is harmful. This point has been mentioned by Chalmers[2] and Stamler.[16] DeMets and Ware[73,74] discuss methods for group sequential designs in which the hypothesis to be tested is one-sided; that is, to test whether the intervention is superior to the control. They proposed retaining the group sequential upper boundaries of Pocock or O'Brien-Fleming for rejection of H_0 while suggesting various forms of a lower boundary which would imply "acceptance" of H_0. One simple approach is to set the lower boundary at an arbitrary value of Z_L such as -1.5 or -2.0. If the test statistic goes below that value, the data may be sufficiently suggestive of a harmful effect to justify terminating the trial. This asymmetric boundary attempts to reflect the behavior or attitude of members of many monitoring committees, which recommend stopping a study once the intervention shows a strong, but nonsignificant, trend in an adverse direction. Work by Gould and Pecore[75] suggests ways for early acceptance of the null hypothesis while incorporating costs as well.

The group sequential methods were developed for clinical trials where results are obtained relatively quickly on a group of $2n$ subjects before another group of $2n$ subjects are entered. The hypothesis being tested could be a comparison of the proportion of successes in each group or the mean level of response. In many trials, however, subjects are entered over a period of time and followed for a relatively long period. Frequently, the primary outcome is time to some event. Instead of adding patients between interim analyses, new events are added. As discussed in Chapter 14, survival analysis methods could be used to compare the experience of the intervention and the control arms. Given their general appeal, it would be desirable to use the group sequential methods in combination with survival analyses.[76] It has been established[77-81] for large studies that the log rank or Mantel-Haenszel statistic[82] can be used. Furthermore, even for small studies, the log rank procedure is still quite robust.[83] The Gehan, or modified Wilcoxon test,[84,85] as defined in Chapter 14 cannot be applied directly to the group sequential procedures. A generalization of the Wilcoxon procedure[86] for survival data, though, is appropriate[78] and the survival methods of analyses can in general terms be applied in group sequential monitoring.

Instead of looking at equal-sized patient groups, the methods strictly require that interim analyses should be done after an additional equal number of events have been observed. Since data monitoring committees usually meet at fixed calendar times, the condition of equal number of events might not be met exactly. However, the methods applied under these circumstances are likely to be approximately correct.

Interim log rank tests in the Beta-Blocker Heart Attack Trial[13] were evaluated using the O'Brien-Fleming group sequential procedure. Seven meetings had been scheduled to review interim data. The trial was designed for a two-sided 5% significance level. These specifications produce the group sequential boundary shown in Figure 15-7. In addition, the interim results of the log rank statistic are also shown for the first six meetings. From the second analysis on, the conventional significance

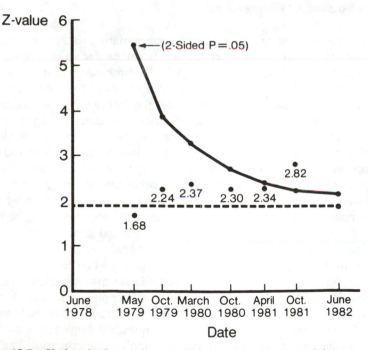

Figure 15-7 Six interim log rank statistics plotted for the time of data monitoring committee meetings with a two-sided O'Brien-Fleming significance level boundary in the Beta-Blocker Heart Attack Trial. Dashed line represents Z = 1.96.[13] Reproduced by permission of Elsevier Science Publishing Co., Inc.

value of 1.96 was exceeded. Nevertheless, the trial was continued. At the sixth meeting, when the O'Brien-Fleming boundary was crossed, a decision was made to terminate the trial with mortality curves as seen in Figure 15-3. It should be emphasized that crossing the boundary was not the only factor in this decision.

While the group sequential methods are a significant advance in data monitoring, the need to specify the number N of planned interim analyses and to require equal numbers of either additional subjects or new events in each analysis has been criticized. A procedure developed by Lan and DeMets[87,88] eliminates these requirements by allowing investigators to determine how they want to "spend" the Type I error during the course of the trial. The error rate spending function guarantees an overall Type I error of size α, at the scheduled end of the trial. Functions are defined which approximate the boundaries described in Figure 15-6. The critical value at a particular decision time is determined by the function and by the number of previous evaluations. However, it is not dependent on future unspecified meetings.

Curtailed Sampling Procedures

During the course of monitoring accumulating data, one question often posed is whether the current trend in the data is so impressive that acceptance or rejection of H_0 is already determined, or at least close to being determined. If the results of the trial are such that the conclusions are "known for certain," no matter what the future outcomes might be, then consideration of early termination is in order. A helpful sports analogy is a baseball team "clinching the pennant" after winning a specific game. At that time it is known for certain who has won and who has not won, regardless of the outcome of the remaining games. Playing the remaining games is done for reasons other than deciding the winner. This idea has been developed for clinical trials and is often referred to as deterministic curtailed sampling. It should be noted that group sequential methods focus on existing data while curtailed sampling in addition considers the data which have not yet been observed.

Alling[89,90] may have introduced this concept when he considered the early stopping question and compared the survival experience in two groups. He used the Wilcoxon test for two samples, a frequently used non-parametric test which ranks survival times and which is the basis for one of the primary survival analysis techniques.[82] Alling's method allows stopping decisions to be based on data available during the trial. The trial would be terminated if future data could not change the final conclusion about the null hypothesis. The method is applicable whether all subjects are entered at the same time or recruitment occurs over a longer period of time. However, when the average time to the event is short relative to the time needed to enroll subjects, the method is of limited value. The repeated testing problem is irrelevant, because any decision is based on what the significance test will be at the end of the study. Therefore, frequent use of this procedure causes no problem with regard to significance level and power.

Many clinical trials with survival time as a response variable have observations which are censored; that is, subjects are followed for some length of time and then at some point, no further information about the subject is known. Halperin and Ware[91] have extended the method of Alling to the case of censored data, using the Wilcoxon rank statistic. With this method, early termination is particularly likely when the null hypothesis is true or when the expected difference between groups is large. The method is shown to be more effective for small sample sizes than for large studies. The Alling approach to early stopping has also been applied to another commonly used test, the Mantel-Haenszel statistic.[92] However, the Wilcoxon statistic appears to have better early stopping properties than the Mantel-Haenszel statistic.

A deterministic curtailed procedure has been developed[93] for comparing the means of two bounded random variables using the two sample t-test. It assumes that the response must be between two values, A and B ($A < B$). An approximate solution is an extreme case approach. First, all the estimated remaining responses in one group are given the maximum favorable outcome (ie, B) and all the remaining responses in the other take on the worst response (ie, A). The statistic is then computed. Next, the responses are assigned in the opposite way and a second statistic is computed. If neither of these two extreme results alters the conclusion, no additional data are necessary for testing the hypothesis. While this deterministic curtailed approach provides an answer to an interesting question, the requirement for absolute certainty results in a very conservative test and allows little opportunity for early termination.

In some clinical trials, the final outcome may not be absolutely certain, but almost so. To use the baseball analogy again, a first place team may not have clinched the pennant but is so many games in front of the second place team that it is highly unlikely that it will not, in fact, end up the winner. Another team may be so far behind that it "realistically" cannot catch up. In clinical trials, this idea is often referred to as stochastic curtailed sampling. It is identical to the concept of conditional power, discussed in the section on extending a trial.

Lan, Simon, and Halperin[94] considered the effect of stochastic curtailed procedures on Type I and Type II error rates. If the null hypothesis, H_0, is tested at time t using a statistic, $S(t)$, then at the scheduled end of a trial at time T, the statistic would be $S(T)$. Two cases are considered. First, suppose a trend in favor of rejecting H_0 is observed at time $t < T$, with intervention doing better than control. One then computes the conditional probability, γ_0, of rejecting H_0 at time T; that is, $S(T) > Z_a$, assuming H_0 to be true and given the current data, $S(t)$. If this probability is sufficiently large, one might argue that the favorable trend is not going to disappear. Second, suppose a negative trend, or data consistent with the null hypothesis of no difference, at some point t. Then, one computes the conditional probability, γ_1, of rejecting H_0 at the end of the trial, time T, given that some alternative is true, for a sample of reasonable alternatives. This essentially asks how large the true effect must be before the current "negative" trend is likely to be reversed. If the probability of a trend reversal is highly unlikely for a realistic range of alternative hypotheses, trial termination might be considered.

Because there is a small probability that the results will change, a slightly greater risk of a Type I or Type II error rate will exist than would be if the trial continued to the scheduled end.[95] However, it has been shown that the Type I error is bounded by α/γ_0 and the Type II error by β/γ_1. For example, if the probability of rejecting the null hypothesis, given the existing data were .85, then the actual Type I error would be no more than .05/.85 or .059, instead of .05.

The Beta-Blocker Heart Attack Trial[13] made considerable use of this approach. As discussed, the interim results were impressive with one year of follow-up still remaining. One question posed was whether the strong favorable trend ($Z = 2.82$) could be lost during that year. The probability of rejecting H_0 at the scheduled end of the trial, given the existing trend, was approximately .90. This meant that the false positive or Type I error was no more than $\alpha/\gamma_0 = .05/.90$ or .056.

Other Approaches

Other techniques for interim analysis of accumulating data have also received attention. These include binomial sampling strategies,[96] decision theoretic models,[97] and likelihood or Bayesian methods.[98-102] The literature on these methods is extensive, but the methods appear to be used infrequently. Cornfield,[98-100] for example, proposed the use of Bayesian methods in clinical trials. These require specifying a prior probability on the possible values of the unknown parameter. The experiment is performed and based on the data obtained, the prior probability is adjusted. If the adjustment is large enough, the investigator may change his opinion (ie, his prior belief). While the Bayesian view is critical of the hypothesis testing methods because of the arbitrariness involved,[101] the Bayesian approach is perhaps hampered mostly by the requirement that the investigator formally specify a prior probability. However, if a person in the decision-making process uses all of the factors and methods discussed in this chapter, a Bayesian approach is involved, although in a very informal way.

Finally, an ad hoc model was developed[103] for stopping accrual of subjects into a clinical trial while follow-up continues for the occurrence of the event of interest. This approach is only applicable if the intervention is given once upon entry into the trial (eg, surgery or vaccination).

REFERENCES

1. Shaw, L.W., and Chalmers, T.C. Ethics in cooperative clinical trials. *Ann NY Acad Sci.* 169:487–495, 1970.

2. Chalmers, T.C. Invited remarks. *Clin Pharmacol Ther.* 25:649–650, 1979.

3. Canner, P.L. Monitoring clinical trial data for evidence of adverse or beneficial treatment effects. Edited by J.P. Boissel and C.R. Klimt. In *Multicenter Controlled Trials: Principles and Problems.* Paris: INSERM, 1979.

4. Robbins, H. Some aspects of sequential design of experiments. *Bull Am Math Soc.* 58:527–535, 1952.

5. Anscombe, F.J. Sequential medical trials. *J Am Stat Assoc.* 58:365–383, 1963.

6. Armitage, P., McPherson, C.K., and Rowe, B.C. Repeated significance tests on accumulating data. *J R Stat Soc.* Series A 132:235–244, 1969.

7. Pocock, S.J. Group sequential methods in the design and analysis of clinical trials. *Biometrika* 64:191–199, 1977.

8. Peto, R., Pike, M.C., Armitage, P. et al. Design and analysis of randomized clinical trials requiring prolonged observations of each patient. I. Introduction and design. *Br J Cancer* 34:585–612, 1976.

9. Tukey, J.W. Some thoughts on clinical trials, especially problems of multiplicity. *Science* 198:679–684, 1977.

10. Canner, P.L. Monitoring of the data for evidence of adverse or beneficial treatment effects. *Controlled Clin Trials* 4:467–483, 1983.

11. Coronary Drug Project Research Group. Practical aspects of decision making in clinical trials: The Coronary Drug Project as a case study. *Controlled Clin Trials* 1:363–376, 1981.

12. DeMets, D.L., Williams, G.W., Brown, B.W., Jr., and the NOTT Research Group. A case report of data monitoring experience: The Nocturnal Oxygen Therapy Trial. *Controlled Clin Trials* 3:113–124, 1982.

13. DeMets, D.L., Hardy, R., Friedman, L.M. and Lan, K.K.G. Statistical aspects of early termination in the Beta-Blocker Heart Attack Trial. *Controlled Clin Trials* 5:362–372, 1984.

14. Fleming, T.R., Green, S.J., and Harrington, D.P. Considerations for monitoring and evaluating treatment effects in clinical trials. *Controlled Clin Trials* 5:55–66, 1984.

15. Klimt, C.R., and Canner, P.L. Terminating a long-term clinical trial. *Clin Pharmacol Ther.* 25:641–646, 1979.

16. Stamler, J. Invited remarks. *Clin Pharmacol Ther.* 25:651–658, 1979.

17. Meier, P. Terminating a trial—the ethical problem. *Clin Pharmacol Ther.* 25:633–640, 1979.

18. Coronary Drug Project Research Group. Influence of adherence to treatment and response of cholesterol on mortality in the Coronary Drug Project. *N Engl J Med.* 303:1038–1041, 1980.

236

19. DeMets, D. Stopping guidelines vs. stopping rules: A practitioner's point of view. *Commun Stat., Theory and Methods* 13:2395–2417, 1984.

20. University Group Diabetes Program. A study of the effects of hypoglycemic agents on vascular complications in patients with adult-onset diabetes. II. Mortality results. *Diabetes* 19(suppl 2):787–830, 1970.

21. University Group Diabetes Program. Effects of hypoglycemic agents on vascular complications in patients with adult-onset diabetes. IV. A preliminary report on phenformin results. *JAMA* 217:777–784, 1971.

22. Report of the Committee for the Assessment of Biometric Aspects of Controlled Trials of Hypoglycemic Agents. *JAMA* 231:583–608, 1975.

23. Kolata, G.B. Controversy over study of diabetes drugs continues for nearly a decade. *Science* 203:986–990, 1979.

24. Coronary Drug Project Research Group. The Coronary Drug Project. Design, methods, and baseline results. *Circulation* 47(Suppl 1):I1–179, 1973.

25. Coronary Drug Project Research Group. The Coronary Drug Project: initial findings leading to modifications of its research protocol. *JAMA* 214:1303–1313, 1970.

26. Coronary Drug Project Research Group. The Coronary Drug Project: findings leading to further modifications of its protocol with respect to dextrothyroxine. *JAMA* 220:996–1008, 1972.

27. Coronary Drug Project Research Group. The Coronary Drug Project: findings leading to discontinuation of the 2.5-mg/day estrogen group. *JAMA* 226:652–657, 1973.

28. Coronary Drug Project Research Group. Clofibrate and niacin in coronary heart disease. *JAMA* 231:360–381, 1975.

29. Nocturnal Oxygen Therapy Group. Continuous or nocturnal oxygen therapy in hypoxemic chronic obstructive lung disease. *Ann Intern Med*. 93:391–398, 1980.

30. Diabetic Retinopathy Study Research Group. Diabetic Retinopathy Study: Report No. 6. design, methods, and baseline results. *Invest Ophthalmol Vis Sci*. 21:149–209, 1981.

31. The Diabetic Retinopathy Study Research Group. Preliminary report on effects of photocoagulation therapy. *Am J Ophthalmol*. 81:383–396, 1976.

32. The Diabetic Retinopathy Study Research Group. Photocoagulation treatment of proliferative diabetic retinopathy: the second report of the Diabetic Retinopathy Study findings. *Ophthalmol*. 85:82–106, 1978.

33. Ederer, F., Podgor, M.J., and the Diabetic Retinopathy Study Research Group. Assessing possible late treatment effects in stopping a clinical trial early: Diabetic Retinopathy Study Report No. 9. *Controlled Clin Trials* 5:373–381, 1984.

34. Beta-Blocker Heart Attack Trial Research Group. A randomized trial of propranolol in patients with acute myocardial infarction. I. Mortality results. *JAMA* 247:1707–1714, 1982.

35. Hypertension Detection and Follow-up Program Cooperative Group. Five-year findings of the Hypertension Detection and Follow-up Program. Reduction in mortality with high blood pressure, including mild hypertension. *JAMA* 242:2562–2571, 1979.

36. Aspirin Myocardial Infarction Study Research Group. A randomized controlled trial of aspirin in persons recovered from myocardial infarction. *JAMA* 243:661–669, 1980.

37. Multiple Risk Factor Intervention Trial Research Group. Multiple Risk Factor Interventional Trial. Risk factor changes and mortality results. *JAMA* 248:1465–1477, 1982.

38. CASS Principal Investigators and Their Associates. Coronary Artery

Surgery Study (CASS): a randomized trial of coronary artery bypass surgery, survival data. *Circulation* 68:939–950, 1983.

39. Collaborative Group on Antenatal Steroid Therapy. Effect of antenatal dexamethasone administration on the prevention of respiratory distress syndrome. *Am J Obstet Gynecol*. 141:276–287, 1981.

40. The MIAMI Trial Research Group. Metoprolol in acute myocardial infarction (MIAMI): A randomized placebo-controlled trial. *Eur Heart J*. 6:199–214, 1985.

41. Gail, M. Monitoring and stopping clinical trials. Edited by V. Mike and K. Stanley. In *Statistics in Medical Research: Methods and Issues, With Applications in Cancer Research*. New York: John Wiley and Sons, 1982.

42. Gail, M. Nonparametric frequentist methods for monitoring comparative survival studies. Edited by P.R. Krishnaiah and P.K. Sen. In *Handbook of Statistics: Parametric Methods. Vol. IV*. Amsterdam: North Holland Publishing, 1984.

43. DeMets, D., and Lan, G. An overview of sequential methods and their application in clinical trials. *Commun Stat., Theory and Methods* 13:2315–2338, 1984.

44. Armitage, P. *Sequential Medical Trials*. 2nd edition. New York: John Wiley and Sons, 1975.

45. Whitehead, J. *The Design and Analysis of Sequential Clinical Trials*. New York: Halsted Press, 1983.

46. Wald, A. *Sequential Analysis*. New York: John Wiley and Sons, 1947.

47. Silverman, W.A., Agate, F.J. Jr., and Fertig, J.W. A sequential trial of the nonthermal effect of atmospheric humidity on survival of newborn infants of low birth weight. *Pediatrics* 31:719–724, 1963.

48. Armitage, P. Restricted sequential procedures. *Biometrika* 44:9–26, 1957.

49. Truelove, S.C., Watkinson, G., and Draper, G. Comparison of corticosteroid and sulphasalazine therapy in ulcerative colitis. *Br Med J*. 2:1708–1711, 1962.

50. McPherson, C.K., and Armitage, P. Repeated significance tests on accumulating data when the null hypothesis is not true. *J R Stat Soc*. Series A 134:15–25, 1971.

51. Whitehead, J., Jones, D.R., and Ellis, S.H. The analysis of a sequential clinical trial for the comparison of two lung cancer treatments. *Stat Med*. 2:183–190, 1983.

52. Whitehead, J., and Stratton, I. Group sequential clinical trials with triangular continuation regions. *Biometrics* 39:227–236, 1983.

53. Dambrosia, J.M., and Greenhouse, S.W. Early stopping for sequential restricted tests of binomial distributions. *Biometrics* 39:695–710, 1983.

54. Chatterjee, S.K., and Sen, P.K. Nonparametric testing under progressive censoring. *Calcutta Stat Assoc Bull*. 22:13–50, 1973.

55. Muenz, L., Green, S., and Byar, D. Applications of the Mantel-Haenszel statistic to the comparison of survival distributions. *Biometrics* 33:617–626, 1977.

56. Davis, C.E. A two sample Wilcoxon test for progressively censored data. *Commun Statist-Theor Meth*. A7(4):389–398, 1978.

57. Koziol, J., and Petkau, J. Sequential testing of equality of two survival distributions using the modified Savage statistic. *Biometrika* 65:615–623, 1978.

58. Breslow, N., and Haug, C. Sequential comparison of exponential survival curves. *J Am Stat Assoc*. 67:691–697, 1972.

59. Canner, P.L. Monitoring treatment differences in long-term clinical trials. *Biometrics* 33:603–615, 1977.

60. Whitehead, J., and Jones, D. The analysis of sequential clinical trials. *Biometrika* 66:443–452, 1979.

61. Jones, D., and Whitehead, J. Sequential forms of the log rank and modified Wilcoxon tests for censored data. *Biometrika* 66:105–113, 1979.

62. Joe, H., Koziol, J.A., and Petkau, J.A. Comparison of procedures for testing the equality of survival distributions. *Biometrics* 37:327–340, 1981.

63. Nagelkerke, N.J.D., and Hart, A.A.M. The sequential comparison of survival curves. *Biometrika* 67:247–249, 1980.

64. Jones, D.R., and Whitehead, J. Applications of large-sample sequential tests to the analysis of survival data. Edited by H.J. Tagnon and M.J. Staquet. In *Controversies in Cancer: Design of Trials and Treatment*. New York: Masson Publishing, 1979.

65. Haybittle, J.L. Repeated assessment of results in clinical trials of cancer treatment. *Br J Radiol*. 44:793–797, 1971.

66. Pocock, S.J. Size of cancer clinical trials and stopping rules. *Br J Cancer* 38:757–766, 1978.

67. O'Brien, P.C., and Fleming, T.R. A multiple testing procedure for clinical trials. *Biometrics* 35:549–556, 1979.

68. Pocock, S.J. Interim analyses for randomized clinical trials: the group sequential approach. *Biometrics* 38:153–162, 1982.

69. Fairbanks, K. and Madsen, R. *P* values for tests using a repeated significance test design. *Biometrika* 69:69–74, 1982.

70. Freedman, L.S., Lowe, D., and Macaskill, P. Stopping rules for clinical trials. *Stat Med*. 2:167–174, 1983.

71. Jennison, C., and Turnbull, B.W. Repeated confidence intervals for group sequential clinical trials. *Controlled Clin Trials* 5:33–45, 1984.

72. Tsiatis, A.A., Rosnar, G.L., and Mehta, C.R. Exact confidence intervals following a group sequential test. *Biometrics* 40:797–803, 1984.

73. DeMets, D., and Ware, J. Group sequential methods in clinical trials with a one-sided hypothesis. *Biometrika* 67:651–660, 1980.

74. DeMets, D.L., and Ware, J.H. Asymmetric group sequential boundaries for monitoring clinical trials. *Biometrika* 69:661–663, 1982.

75. Gould, A.L., and Pecore, V.J. Group sequential methods for clinical trials allowing early acceptance of H_0 and incorporating costs. *Biometrika* 69:75–80, 1982.

76. Seigel, D., and Milton, R.C. Further results on a multiple-testing procedure for clinical trials. *Biometrics* 39:921–928, 1983.

77. Tsiatis, A.A. The asymptotic joint distribution of the efficient scores tests for the proportional hazards model calculated over time. *Biometrika* 68:311–315, 1981.

78. Tsiatis, A.A. Repeated significance testing for a general class of statistics used in censored survival analysis. *J Am Stat Assoc*. 77:855–861, 1982.

79. Sellke, T., and Siegmund, D. Sequential analysis of the proportional hazards model. *Biometrika* 70:315–326, 1983.

80. Tsiatis, A.A. Group sequential methods for survival analysis with staggered entry. In *Survival Analysis. Monograph Series 2*. Edited by R. Johnson and J. Crowley. Hayward, California: IMS Lecture Notes, pp. 257–268, 1982.

81. Harrington, D.P., Fleming, T.R., and Green, S.J. Procedures for serial testing in censored survival data. In *Survival Analysis. Monograph Series 2*. Edited by R. Johnson and J. Crowley. Hayward, California: IMS Lecture Notes, pp. 269–286, 1982.

82. Mantel, N. Evaluation of survival data and two new rank order statistics arising in its consideration. *Cancer Chemother Rep*. 50:163–170, 1966.

83. Gail, M.H., DeMets, D.L., and Slud, E.V. Simulation studies on increments of the two-sample log rank score test for survival data, with application to group sequential boundaries. Edited by R. Johnson and J. Crowley. In *Survival Analysis. Monograph Series 2*. Hayward, California: IMS Lecture Notes, pp. 287–301, 1982.

84. Slud, E., and Wei, L.J. Two-sample repeated significance tests based on the modified Wilcoxon statistic. *J Am Stat Assoc*. 77:862–868, 1982.

85. Gehan, E. A generalized Wilcoxon test for comparing arbitrarily singly-censored samples. *Biometrika* 52:203–224, 1965.

86. Peto, R., and Peto, J. Asymptotically efficient rank invariant test procedures. *J R Stat Soc*. Series A 135:185–206, 1972.

87. Lan, K.K.G., and DeMets, D.L. Discrete sequential boundaries for clinical trials. *Biometrika* 70:659–663, 1983.

88. Lan, K.K.G., DeMets, D.L., and Halperin, M. More flexible sequential and non-sequential designs in long-term clinical trials. *Commun Stat., Theory and Method* 13:2339–2354, 1984.

89. Alling, D.R. Early decision in the Wilcoxon two sample test. *J Am Stat Assoc*. 58:713–720, 1963.

90. Alling, D.W. Closed sequential tests for binomial probabilities. *Biometrika* 53:73–84, 1966.

91. Halperin, M., and Ware, J. Early decision in a censored Wilcoxon two-sample test for accumulating survival data. *J Am Stat Assoc*. 69:414–422, 1974.

92. Verter, J. Early decision in a Mantel-Haenszel statistic for accumulating survival data. PhD Thesis, University of North Carolina, 1979.

93. DeMets, D.L., and Halperin, M. Early stopping in the two-sample problem for bounded random variables. *Controlled Clin Trials* 3:1–11, 1982.

94. Lan, K.K.G., Simon, R., and Halperin, M. Stochastically curtailed tests in long-term clinical trials. *Commun Stat., Sequential Analysis*, 1:207–219, 1982.

95. Halperin, M., Lan, K.K.G., Ware, J.H., et al. An aid to data monitoring in long-term clinical trials. *Controlled Clin Trials* 3:311–323, 1982.

96. Simon, R., Weiss, G.H., and Hoel, D.G. Sequential analysis of binomial clinical trials. *Biometrika* 62:195–200, 1975.

97. Colton, T. A model for selecting one of two medical treatments. *J Am Stat Assoc*. 58:388–400, 1963.

98. Cornfield, J. Sequential trials, sequential analysis and the likelihood principle. *Am Statistician* 20:18–23, April 1966.

99. Cornfield, J. A Bayesian test of some classical hypotheses – With applications to sequential clinical trials. *J Am Stat Assoc*. 61:577–594, 1966.

100. Cornfield, J. Recent methodological contributions to clinical trials. *Am J Epidemiol*. 104:408–421, 1976.

101. Berry, D.A. Interim analyses in clinical trials: Classical vs. ad hoc vs. Bayesian approaches. Technical Report No. 418, University of Minnesota School of Statistics, 1983.

102. Lachin, J.M. Sequential clinical trials for normal variates using interval composite hypotheses. *Biometrics* 37:87–101, 1981.

103. Rubinstein, L.V. and Gail, M.H. Monitoring rules for stopping accrual in comparative survival studies. *Controlled Clin Trials* 3:325–343, 1982.

CHAPTER 16

Issues in Data Analysis

The analysis of data obtained from a clinical trial represents the outcome of the planning and implementation already described. It allows one to test to the primary and secondary questions addressed by the clinical trial and provides for the generation of new hypotheses. Data analysis is sometimes viewed as simple and straightforward, requiring little time, effort, or expense. However, careful analysis usually requires a major investment in all three. It must be done with as much care and concern as any of the design or data-gathering aspects. Furthermore, inappropriate statistical analyses can result in misleading conclusions and impair the credibility of the trial.

Several introductory textbooks of statistics[1-6] provide excellent descriptions for many basic methods of analysis. Chapter 14 presents essentials for analysis of survival data, since they are frequently of interest in clinical trials and are not covered in most introductory statistics texts. This chapter will focus on some issues in the analysis of data which seem to cause confusion in the medical research community. Some of the proposed solutions are straightforward; others are judgmental. They reflect a point of view developed in several collaborative efforts supported by the National Heart, Lung, and Blood Institute over the past decade. Whereas some[7,8] have taken similar positions, others[9,10] have opposing views on several issues.

FUNDAMENTAL POINT

Excluding randomized subjects from analysis and subgrouping on the basis of outcome or response variables can lead to biased results. This bias can be of unknown magnitude or direction.

WHICH SUBJECTS SHOULD BE ANALYZED?

The issue of which subjects are to be included in the data analysis often arises in clinical trials. Although a laboratory study may have carefully regulated experimental conditions, even the best designed and managed clinical trial cannot be perfectly implemented. Response variable data may be missing, the protocol may not be completely complied with, and some subjects, in retrospect, will not have met the entrance criteria. Some investigators prefer to remove from the analysis subjects who do not fit the design or do not follow the protocol perfectly. Conversely, others believe that once a subject is randomized, that subject should always be followed and included in the analysis. The rationale for each of these positions is considered in the following pages. This chapter has adopted, in part, the terminology used by Peto and colleagues[7] to classify subjects according to the nature and extent of their participation.

Exclusions are people who are screened as potential subjects for a randomized trial but who do not meet all of the entry criteria and, therefore, are not enrolled. Reasons for exclusion might be related to age, severity of disease, refusal to participate, or any of numerous other determinants evaluated before allocation of a subject to a study group (Chapter 3). Because such people are not enrolled in the trial, they do not bias the randomization or the intervention-control group comparison. Exclusions do, however, influence interpretation of the results of the clinical trial. In some circumstances, follow-up of excluded people can be helpful in determining to what extent the results can be generalized. If the event rate in the control group is considerably lower than anticipated, an investigator may determine whether most high risk people were excluded or whether he was incorrect in his initial assumption.

Withdrawals are subjects who have been enrolled but are deliberately not counted in the analysis. As the fundamental point states, omitting subjects from analyses can bias the results of the study.[11] If subjects are withdrawn, the burden rests with the investigator to convince the scientific community that the analysis has not been biased. However, this can be a difficult task, because no one can be sure that subjects were not differentially withdrawn from the study groups. Differential withdrawal can exist even if the number of omitted subjects is the same in each group, since the reasons for withdrawal in each group may be different. The subjects remaining in the trial may not be comparable, undermining one of the reasons for randomization.

REASONS FOR WITHDRAWAL

Commonly cited reasons for withdrawal of subjects from the analysis are ineligibility, noncompliance, poor quality data, and occurrence of competing events.

Ineligibility The first reason for withdrawal involves subjects who did not meet the entry criteria, a protocol violation unknown at the time of enrollment. Admitting unqualified subjects could occur by a simple clerical error, a laboratory error, a misinterpretation, or a misclassification. Clerical mistakes such as listing wrong sex or age may be obvious. Other errors can arise from differing interpretation of diagnostic studies such as electrocardiograms, x-rays, or biopsies.

Withdrawals for ineligibility can involve a relatively large number of subjects. In a trial by the Canadian Cooperative Study Group,[12] 64 of the 649 enrolled subjects (10%) were later found to have been ineligible. In this four-armed study, the numbers of ineligible patients in the study groups ranged from 10 to 25. The reasons for the ineligibility of these 64 subjects were not reported, nor was the outcome experience in this group given.

A study design may require enrollment within a defined time period following a qualifying episode. Because of this time constraint, data concerning a subject's eligibility might not be available or confirmed when the decision is made to enroll him. For example, the Beta-Blocker Heart Attack Trial looked at two-year mortality in people administered a beta-blocking drug during hospitalization for an acute myocardial infarction.[13] Because of known variability in interpretation, the protocol required that the diagnostic electrocardiograms be read by a central unit. However, this verification took several weeks to accomplish. Local institutions, therefore, interpreted the electrocardiograms and decided whether the patient met the necessary criteria for inclusion. Almost 9% of the enrolled subjects did not have their myocardial infarction confirmed, and were "incorrectly" randomized. The question then arises: Should the subjects be kept in the trial and included in the analysis of the response variable data? The Beta-Blocker Heart Attack Trial protocol required follow-up and analysis of all randomized subjects. In this case, it made no important difference. The observed benefits from the intervention were similar in those eligible as well as in those ineligible.

A more complicated situation occurs when the data needed for enrollment cannot be obtained until hours or days have passed, yet the study design requires initiation of intervention before then. For instance, in the Multicenter Investigation for the Limitation of Infarct Size,[14] propranolol, hyaluronidase, or placebo was administered shortly after subjects were admitted to the hospital with possible acute myocardial infarctions. In some, the diagnosis of myocardial infarction might not have been confirmed until after electrocardiographic and serum enzyme changes had been monitored for several days. Such subjects were, therefore, randomized on the basis of preliminary diagnoses of infarction. Subsequent review of the course of illness may not have supported the diagnosis. Another example of this problem involves a study of pregnant women who were likely to deliver prematurely and, therefore, would have children who were at a higher than usual risk of being born

244

with respiratory distress syndrome.[15] Corticosteroids administered to the mother prior to delivery were hypothesized to protect the premature child from developing respiratory distress syndrome. Although, at the time of the mother's randomization to either intervention or control groups, the investigator could not be sure that the delivery would be premature, he needed to make a decision whether to enroll the mother into the study.

To complicate matters still further, the intervention given to a subject can affect or change the entry diagnosis. For example, in the abovementioned study to limit infarct size, some patients without a myocardial infarction were randomized because of the need to begin intervention before the diagnosis was confirmed. Moreover, if the interventions succeeded in limiting infarct size, they could have affected the electrocardiogram and serum enzyme levels. Subjects in the intervention groups with a small myocardial infarction may have shown reduced serum enzyme values and appeared not to have had an infarction. Thus, they would not seem to have met the entry criteria. However, this situation could not exist in the placebo control group. If the investigators had withdrawn subjects who did not meet the study criteria for a myocardial infarction, they would have withdrawn more subjects from the intervention groups (those with no infarction plus those with small infarction) than from the control group (those with no infarction). This could have produced a bias in later comparisons. On the other hand, it could be assumed that a similar number of truly ineligible subjects were randomized to the intervention groups and to the control group. In order to maintain comparability, the investigators might have decided to withdraw the same number of subjects from each group. The ineligible subjects in the control group could have been readily identified. However, the subjects in the intervention groups who were truly ineligible had to be distinguished from those made to appear ineligible by the effects of the interventions. This would have been difficult, if not impossible. In the Multicenter Investigation for the Limitation of Infarct Size, all randomized subjects were kept in the analysis.[14]

An example of possible bias because of withdrawal of ineligible subjects is found in the Anturane Reinfarction Trial, which compared sulfinpyrazone with placebo in subjects who had recently suffered a myocardial infarction.[16] As seen in Table 16-1, of 1629 randomized subjects (813 to sulfinpyrazone, 816 to placebo), 71 were subsequently found to be ineligible. Thirty-eight had been assigned to sulfinpyrazone and 33 to placebo. Despite relatively clear definitions of eligibility and comparable numbers of subjects withdrawn, mortality among these ineligible subjects was 26.3% in the sulfinpyrazone group (10 of 38) and 12.1% in the placebo group (4 of 33).[17] The eligible placebo group subjects had a mortality of 10.9%, similar to the 12.1% seen among the ineligible sub-

Table 16-1
Mortality by Study Group and Eligibility Status
in the Anturane Reinfarction Trial

	Random-ized	Percent Mortality	Ineligible	Percent Mortality	Eligible	Percent Mortality
Sulfinpyrazone	813	9.1	38	26.3	775	8.3
Placebo	816	10.9	33	12.1	783	10.9

jects. In contrast, the eligible subjects on sulfinpyrazone had a mortality of 8.3%, less than one-third that of the ineligible subjects. Including all 1629 subjects in the analysis gave 9.1% mortality in the sulfinpyrazone group, and 10.9% mortality in the placebo group ($P = .20$). Withdrawing the 71 ineligible subjects (and 14 deaths) gave an almost significant $P = .07$.

Stimulated by criticisms of the study, the investigators initiated a reevaluation of the Anturane Reinfarction Trial results. An independent group of reviewers examined all reports of deaths in the trial.[18] Instead of 14 deceased subjects who were ineligible, it found 19; 12 in the sulfinpyrazone group and 7 in the placebo group. Thus, supposedly clear criteria for ineligibility can be judged differently.

Three trial design policies that relate to withdrawals because of entry criteria violations have been discussed by Peto et al.[7] The first policy is not to enroll subjects until all the diagnostic tests have been confirmed and all the entry criteria have been carefully checked. Once enrollment takes place, no withdrawals are allowed. This option may result in loss of some eligible subjects. For some studies, such as the one on limiting infarct size, this policy cannot be applied because firm diagnoses cannot be ascertained prior to the time when intervention has to be initiated.

The second policy is to enroll marginal or unconfirmed cases and later withdraw those subjects who are proven to have been misdiagnosed. This would be allowed, however, only if the decision to withdraw is based on data collected before enrollment. The process of deciding upon withdrawal of a subject from a study group should be done blinded with respect to the subject's outcome and group assignment.

A third policy is to enroll some subjects with unconfirmed diagnoses and to allow no withdrawals. What makes this procedure valid is that the investigator compares two randomized groups. However, this policy is conservative because each group contains some subjects who might not be able to benefit from intervention. Thus, the overall trial may have less power to detect differences of interest.

A modification to these three policies is recommended. Every effort should be made to establish the eligibility of subjects for entry into the study within the constraints of the design. No withdrawals should be allowed and the analyses should include all subjects enrolled. Subgroup analyses on the basis of baseline eligibility data might also be performed. If the analyses of data from all enrolled subjects and from subgroups agree, the interpretation of the results, at least with respect to subject eligibility, is clear. If the results differ, however, the investigator must be very cautious in his interpretation. In general, he should emphasize the analysis with all the enrolled subjects because that analysis is always valid.

Any policy on withdrawals should be stated in the study protocol before the start of the study. The actual decision to withdraw specific subjects should be done without knowledge of the study group, ideally by someone not directly involved in the trial. Of special concern is withdrawal based on review of selected cases, particularly if the decision rests on a subjective interpretation. Even in double blind trials, blinding may not be perfect, and the investigator may supply information differentially depending upon study group and health status. Therefore, withdrawal should be done early in the course of follow-up, before a response variable has occurred, and with a minimum of data exchange between the investigator and the person making the decision to withdraw the subject. This withdrawal approach does not preclude a later challenge by readers of the report, on the basis of potential bias. It should, however, remove the concern that the withdrawal policy was dependent on the outcome of the trial. The withdrawal rules should not be based on knowledge of study results. Even when these guidelines are adhered to, if the number of entry criteria violations are substantially different in the study groups, or if the event rates in the withdrawn subjects are different between the groups, the question will certainly be raised whether bias played a role in the decision to withdraw a subject.

Noncompliance The second reason that subjects are withdrawn is noncompliance to the prescribed intervention or control regimen. Noncompliance refers to dropouts and dropins (Chapter 13). The decision not to comply with the protocol intervention may be made by the subject, his primary care physician, or the trial investigator. Noncompliance may be due to adverse effects of the intervention or control, loss of subject interest, changes in the underlying condition of a subject, or a variety of other reasons.

Withdrawal from analysis of subjects who do not comply with the intervention regimens specified in the study design is often proposed. The motivation for withdrawal of noncompliant subjects is that the trial is not a fair test of the intervention with these subjects included. For example, there may be a few subjects in the intervention group who took little or no therapy. If subjects do not take their medication, they certainly cannot benefit from it. There could also be subjects in the control group who frequently receive the study medication. The intervention and control groups are thus "contaminated." Proponents of withdrawal of non-complying subjects argue that removal of these subjects keeps the trial closer to what was intended; that is, a comparison of optimal intervention versus control. The impact of noncompliance on the trial is that any observed benefits of the intervention, as compared to the control, will be reduced, thus making the trial less powerful than it potentially could have been.

A policy of withdrawal from analysis because of subject non-compliance can lead to bias. The overwhelming reason is that subject compliance to a protocol may be related to the intervention. In other words, there may be an interaction between compliance and intervention. Certainly, if noncompliance is greater in one group than another, then withdrawal of noncompliant subjects could lead to bias. Even if the frequency of noncompliance is the same for the intervention and control groups, the reasons for noncompliance in each group may differ and may involve different types of subjects. The concern would always be whether the same type of subject had been withdrawn in the same proportion from each group or whether an imbalance had been created. Of course, an investigator could probably neither confirm nor refute the possibility of bias.

The Coronary Drug Project evaluated several lipid-lowering drugs in people several years after a myocardial infarction. In subjects on one of the drugs, clofibrate, total five-year mortality was 18.2%, as compared with 19.4% in control subjects.[19] Among the clofibrate subjects, those who had at least 80% compliance to therapy had a mortality of 15%, whereas the poor compliers had a mortality of 24.6% (Table 16-2). This seeming benefit from taking clofibrate was, unfortunately, mirrored in the group taking placebo, 15.1% vs. 28.2%. A similar pattern (Table 16-3) was noted in the Aspirin Myocardial Infarction Study.[20] Overall, no difference in mortality was seen between the aspirin-treated group (10.9%) and the placebo-treated group (9.7%). Good compliers to aspirin had a mortality of 6.1%; poor compliers had a mortality of 21.9%. In the placebo group, the rates were 5.1% and 22%.

A trial of antibiotic prophylaxis in cancer patients also demonstrated

Table 16-2
Percent Mortality by Study Group and Level of Compliance in the Coronary Drug Project

	Overall	Drug Compliance	
		≥80%	<80%
Clofibrate	18.2	15.0	24.6
Placebo	19.4	15.1	28.2

Table 16-3
Percent Mortality by Study Group and Degree of
Compliance in the Aspirin Myocardial Infarction Study

	Overall	Good Compliers	Poor Compliers
Aspirin	10.9	6.1	21.9
Placebo	9.7	5.1	22.0

a relationship between compliance and benefit in both the intervention and placebo groups.[21] Among the subjects assigned to intervention, efficacy in reducing fever or infection was 82% in excellent compliers, 64% in good compliers, and 31% in poor compliers. Among the placebo subjects, the corresponding figures were 68%, 56%, and 0%.

One of the interventions in the Coronary Drug Project, high dose estrogen, produced a somewhat different pattern. As discussed by Canner,[22] the mortality of the subjects in the placebo group who complied with the therapy, that is, who took 80% or more of the protocol dose, was 4.8%. The mortality of the placebo subjects who did not comply was 9.9%. Both compliers and noncompliers to estrogen had similar mortality: 6.3% and 6.1%. The finding that noncompliers to placebo had a different outcome from noncompliers to active intervention (9.9% vs. 6.1%) could theoretically lead one to conclude that "it is beneficial to be randomized to the estrogen group and not take the drug."[22]

A third pattern is noted in a three-arm trial comparing two beta-blocking drugs, propranolol and atenolol, with placebo.[23] Approximately equal numbers of subjects in each group stopped taking their medication. In the placebo group, compliers and noncompliers had similar mortality: 11.2% and 12.5%, respectively. Noncompliers to the interventions, however, had death rates several times greater than did the compliers: 15.9% to 3.4% in those on propranolol and 17.6% to 2.6% in those on atenolol. Thus, even though the numbers of noncompliers were similar, their characteristics were obviously different.

Detre and Peduzzi have argued that, although as a general rule noncompliant subjects should be analyzed according to the study group to which they were assigned, there can be exceptions. They present an example from the VA coronary bypass surgery trial.[24] In that trial, a number of subjects assigned to medical intervention crossed over to surgery. Contrary to expectation, these subjects were at similar risk of having an event, after accounting for a variety of baseline factors, as those who did not crossover. Therefore, the authors argued that the noncompliers can be censored at the time of crossover. This may be true, but, as seen in the Coronary Drug Project,[19] adjustment for known variables does not always account for the observed response. The differences in mortality between compliers and noncompliers remained even after adjustment. Thus, other unmeasured variables were of critical importance. The same may hold true for the VA trial.

It has been claimed[10] that if rules for withdrawing subjects are specified in advance, withdrawals for noncompliance reasons are legitimate. However, the potential for bias cannot be avoided simply because the investigator states, ahead of time, the intention to withdraw subjects. This is true even if the investigator is blinded to the group assignment of a subject at the time of withdrawal. Subjects were not withdrawn from the analyses in the above examples. However, had a rule allowing withdrawal of subjects with poor compliance been specified in advance, the type of subjects withdrawn would have been different in the groups. This could have resulted in the analysis of noncomparable groups of compliers. Unfortunately, as noted, the patterns of possible bias can vary. Neither the magnitude nor direction of that bias is easily assessed or compensated for in analysis.

Compliance is also a response to the intervention. If the compliance by subjects to an intervention is poor compared to that by subjects to control, even after the best efforts by the investigators, widespread use of this therapy in the study population may not be feasible. An intervention may be effective, but if it cannot be tolerated by a large portion of the subjects, then that fact is also important.

It is therefore recommended that no subjects be withdrawn because of compliance reasons. The price an investigator must pay for this policy is possibly to reduce the power of the study because some subjects who are included may not be on optimal intervention. For limited or moderate noncompliance, one can compensate for this reduced power by increasing the sample size. Increasing the sample size is costly and undesirable, but the alternative action may make the comparison between the intervention and control biased. This is even more unsatisfactory.

Poor quality data   Subjects may be withdrawn from a trial because the data on them are found to be of poor quality, the extreme being missing data.

In long-term trials subjects may be lost to follow-up. In this situation, the status of the subject with regard to any response variable cannot be determined. If mortality is the primary response variable and if the subject fails to return to the clinic, his survival status may still be obtained. If a death has occurred, the date of death can be ascertained. In the Coronary Drug Project[25] where survival experience over 60 months was the primary response variable, four of 5011 subjects were lost to follow-up (one in a placebo group, three in one treatment group, and none in another treatment group). The Lipid Research Clinics Coronary Primary Prevention Trial[26] followed over 3800 subjects for an average of 7.4 years, and was able to assess vital status on all. Obtaining such low loss to follow-up rates, however, required special efforts.

An investigator may not be able to obtain information on other response variables. For example, if a subject is to have blood pressure measured at the last follow-up visit 12 months after randomization and the subject does not appear, that 12-month blood pressure can never be retrieved. Even if the subject is contacted later, the later measurement does not truly represent the 12-month blood pressure. In some situations, substitutions may be permitted, but, in general, this will not be a satisfactory solution. An investigator needs to make every effort to have subjects come in for their scheduled visits in order to keep losses to follow-up at a minimum.

If the rate of loss to follow-up is different in each of two study groups, there could be a problem in the analysis of the data. A bias could be introduced if the loss is related to the intervention. For example, subjects who are taking a particular new drug which has some adverse effects may not be doing well. They may not come back for their clinic visit and yet may have a subsequent unobserved event. These losses to follow-up would probably not be the same in the control group. In this situation, there is a bias favoring the new drug. Of course, in reality one cannot be sure whether such a bias exists, but one must be careful that such a possibility does not affect the outcome. Even if the rates of loss to follow-up are the same in each study group, the investigator should not be too quick to dismiss the possibility of bias. The recurrent argument that different mechanisms may be involved also applies. The subjects who are lost in each group may have been quite different in their prognoses and eventual outcomes.

Survival analysis methods, which are discussed in Chapter 14, consider time from initiation of intervention to response. These methods can use the experience of subjects up to the time of loss to follow-up if certain conditions are met. For continuous variables, the analytical methods are not as carefully worked out. For example, an investigator may be measuring change in a variable over time. If a subject is lost to follow-up, the observed rate of change may be used to estimate the missing data. In other situations, an investigator might elect to rank the observed response variables and assign a rank to those subjects with missing data. He could assign the highest rank to all such subjects in the intervention group, while assigning the lowest ranks to those in the control group. Then he could analyze the ranked response variable data. Next, the way in which the investigator assigned highest and lowest ranks could be reversed and the analysis repeated. If the results are consistent, there would be some assurance that the lost information did not have any effect on the overall study conclusions. If the results are inconsistent, the investigator would have to be more cautious in his interpretation. Since losses to follow-up can lead to bias and can complicate the analyses, every effort should be made to keep them to a minimum.

Data may not be totally missing. They may, however, look erroneous or inconsistent with other information. It is tempting to withdraw subjects with such data from analysis. There is always the possibility, however, that the erroneous data are associated with subjects in a particular study group. Therefore, any withdrawal of subjects needs to be done carefully, with adequately documented reasons and procedures.

An outlier is an extreme value significantly different from the remaining values. The concern is, whether extreme values in the sample should be excluded from the analysis. This question may apply to a laboratory result, to the data from one of several wards in a hospital or from a clinic in a multicenter trial. Removing outliers is not recommended unless the data can be clearly shown to be erroneous. Even though a value may be an outlier, it could be correct, indicating that on occasions an extreme result is possible. This fact could be very important and should not be ignored by eliminating the outlier. Kruskal[27] suggests carrying out an analysis with and without the "wild observation." If the conclusions vary depending on whether the outlier values are included or excluded, one should view any conclusions cautiously. Procedures for detecting extreme observations have been discussed,[28-32] and the publications cited can be consulted for further details.

An interesting example given by Canner et al[31] concerns the Coronary Drug Project. The authors plotted the distributions of four response variables for each of the 53 clinics in that multicenter trial. Using total mortality as the response variable, no clinics were outlying. When nonfatal myocardial infarction was the outcome, only one clinic was an outlier. With congestive heart failure and angina pectoris, response variables which are probably less well defined, there were nine and eight outlying clinics, respectively.

Competing events Competing events preclude the assessment of the primary response variable. They can reduce the power of the trial by decreasing the number of subjects available for follow-up. If the intervention can affect the competing event, there is also the risk of bias. In some clinical trials, the primary response variable may be cause-specific mortality, such as death due to myocardial infarction or sudden death, rather than total mortality. The reason for using cause-specific death as a response variable is that a therapy often has specific mechanisms of action which are effective against a disease or condition. In this situation, measuring death from all causes — most of which are not likely to be affected by the intervention — can "dilute" the results. For example, a study drug may be antiarrhythmic and thus sudden cardiac death might be the selected response variable. Other causes of death would be competing events.

Even if the response variable is not cause-specific mortality, death may be a factor in the analysis. This is particularly a problem in long-term trials in the elderly or high risk populations. If a subject dies, the measurements which would have been obtained are missing. Analysis of nonfatal response variable data on surviving subjects has the potential for bias, especially if the mortality rates are different in the two groups.

In a study in which a cause-specific death is the primary response variable, deaths from other causes are treated statistically as though the subjects were lost to follow-up from the time of death (Chapter 14) and these deaths are not counted in the analysis. In this situation, the analysis, however, must go beyond merely examining the primary response variable. An intervention may or may not be effective in treating the condition of interest but could be harmful in other respects. Therefore, total mortality should be considered as well as the cause-specific fatal event. Similar considerations need to be made when death occurs in studies using nonfatal primary response variables. This argument is more than theoretical. The Coronary Primary Prevention Trial[26] demonstrated a reduction in coronary heart disease mortality and morbidity in subjects with elevated cholesterol who were taking the lipid-lowering drug, cholestyramine. Total mortality, however, was almost identical in the intervention and placebo control groups. This was due to a larger number of deaths from violent causes in the intervention group. As there was no good explanation for this, the investigators attributed it to chance. This may well be the case. However, the same pattern of increased noncardiac mortality appears in a number of trials of lipid-lowering regimens, including diet.[35]

No completely satisfactory solution exists for handling competing events. At the very least, the investigator should report all major outcome categories; for example, total mortality, as well as cause-specific mortality and morbid events.

COVARIATE ADJUSTMENT

The goal in a clinical trial is to have groups of subjects that are comparable except for the intervention being studied. Even if randomization is used, all of the prognostic factors may not be perfectly balanced, especially in smaller studies. Even if no prognostic factors are significantly imbalanced in the statistical sense, an investigator may, nevertheless, observe that one or more factors favor one of the groups. In either case, covariate adjustment can be used in the analysis to minimize the effect of the differences.

Adjustment also reduces the variance in the test statistic. If the covariates are highly correlated with outcome, this can produce a more sensitive analysis. The specific adjustment procedure depends on the type of covariate being adjusted for and the type of response variable being analyzed. If a covariate is discrete, or if a continuous variable is converted into intervals and made discrete, the analysis is sometimes referred to as "stratified." A stratified analysis, in general terms, means that the study subjects are subdivided into smaller, more homogeneous groups, or strata. A comparison is made within each stratum and then averaged over all strata to achieve a summary result for the response variable. This summary result is adjusted for group imbalances in the discrete covariate. If a response variable is discrete, such as the occurrence of an event, the stratified analysis might take the form of a Mantel-Haenszel statistic,[33] described briefly in the Appendix to this chapter. Bishop et al[34] discuss analysis of discrete data in detail.

If the response variable is continuous, the stratified analysis is referred to as analysis of covariance. This uses a model which, typically, is linear in the covariates. A simple example for a response Y and covariate X would be $Y = \alpha_j + \beta(X - \mu) + \text{error}$ where β is a coefficient representing the importance of the covariate X and is assumed to be the same in each group, μ is the mean value of X, and α is a parameter for the contribution of the overall response variable j^{th} group (eg, $j = 1$ or 2). The basic idea is to adjust the response variable Y for any differences in the covariate X between the two groups. Under appropriate assumptions, the advantage of this method is that the continuous covariate X does not have to be divided into categories. Further details concerning the use of this methodology can be found in statistics textbooks.[1,3] If time to an event is the primary response variable, then survival analysis methods are used (Chapter 14). These methods allow for adjustments of discrete or continuous covariates.

Regardless of the adjustment procedure, covariates should be measured at baseline. Except for certain factors such as age, sex, or race, any variables that are evaluated after initiation of intervention should be considered as response variables. Group comparisons of the primary response variable, adjusted for other response variables, are discouraged. Interpretation of such analyses are difficult because group comparability may be lost.

As discussed earlier, compliance is also a response variable. Adjusting for compliance can lead to misinterpretation of results. Sometimes, adjustment in analysis for compliance may not be obvious. In a trial of clofibrate,[36] the authors reported that those subjects who had the largest reduction in serum cholesterol had the greatest clinical improvement. However, reduction in cholesterol is probably highly correlated with compliance to the intervention regimen. Since compliers in one group may be different from compliers in another group, analyses which adjust for compliance can be biased.

This issue was addressed in the Coronary Drug Project.[19] Adjusted for baseline factors, the five-year mortality was 18.8% in the clofibrate group (N = 997) and 20.2% in the placebo group (N = 2535), an insignificant difference. For subjects with baseline serum cholesterol greater than or equal to 250 mg/dl, the mortality was 17.5% and 20.6% in the clofibrate and placebo groups, respectively. No difference in mortality between the groups was noted for subjects with baseline cholesterol of less than 250 mg/dl (20.0% vs. 19.9%).

Those subjects with lower baseline cholesterol in the clofibrate group who had a reduction in cholesterol during the trial had a 16.0% mortality, as opposed to a 25.5% mortality for those with a rise in cholesterol (Table 16-4). This would fit the hypothesis that lowering cholesterol is beneficial. However, in those subjects with high baseline cholesterol, the situation was reversed. An 18.1% mortality was seen in those who had a fall in cholesterol, and a 15.5% mortality was noted in those who had a rise in cholesterol. The best outcome, therefore, appeared to be in subjects on clofibrate whose low baseline cholesterol dropped or whose high baseline cholesterol increased. Lack of knowledge or understanding about why some people comply or respond to intervention while others do not makes this sort of analysis vulnerable to misinterpretation and bias.

Clinical trials of cancer treatment commonly analyze results by comparing responders with nonresponders. That is, those who go into remission or have a reduction in tumor size are compared with those who do not. One survey indicates that such analyses have been done in at least 20% of published reports.[37] The authors of the survey argue that statistical problems, due to lack of random assignment, and methodological problems, due both to classification of response and inherent differences between responders and nonresponders, can occur. These will often lead to misinterpretations of the study results. Anderson et al[38] provide an example of the bias that can accompany such analyses. They point out that subjects "who eventually become responders must survive

Table 16-4
Percent Five-Year Mortality in the Clofibrate Group, by Baseline Cholesterol and Change in Cholesterol in the Coronary Drug Project[19]

	Baseline Cholesterol	
	<250 mg/dl	≥250 mg/dl
Total	20.0	17.5
Fall in cholesterol	16.0	18.1
Rise in cholesterol	25.5	15.5

long enough to be evaluated as responders." This factor can invalidate some statistical tests that compare responders with nonresponders. The authors present two statistical tests that avoid bias. They note, though, that even if the tests indicate a significant difference in survival between responders and nonresponders, it cannot be concluded that increased survival is due to tumor response. Thus, aggressive intervention, which may be associated with better response, cannot be assumed to be better than less intensive intervention, which may be associated with poorer response. Anderson and colleagues state that only a truly randomized comparison can say which intervention method is preferable. What is unsaid, and illustrated by the Coronary Drug Project examples, is that even comparison of good responders in the intervention group with good responders in the control can be misleading, because there may be different reasons for good response.

The Cox proportional hazards regression model for the analysis of survival data (Chapter 14) allows for covariates in the regression to vary with time.[39] This has been suggested as a way to adjust for factors such as compliance and level of response. It should be pointed out that this and simple regression models are vulnerable to the same biases described earlier in this chapter.

The issue of stratification was first raised in the discussion of randomization (Chapter 5). For large studies, the recommendation was that stratified randomization is usually unnecessary because overall balance would nearly always be achieved and that stratification would be possible in the analysis. For smaller studies, baseline adaptive methods could be considered but the analysis should include the covariates used in the randomization. In a strict sense, analysis should always be stratified if stratification was used in the randomization. In such cases, the adjusted analysis should include not only those covariates found to be different between the groups, but also those stratified during randomization. Of course, if no stratification is done at randomization, the final analysis is less complicated since it would involve only those covariates that turn out to be imbalanced or to be of special interest.

In multicenter trials the randomization should be stratified by clinic. The analysis of such a study should, therefore, incorporate the clinic as a stratification variable. Furthermore, the randomization should be blocked in order to achieve balance over time in the number of subjects randomized to each group. These "blocks" are also strata and, ideally, should be included in the analysis as a covariate. However, there could be a large number of strata, since there may be many clinics and the blocking factor within any clinic is usually anywhere from four to eight subjects. Use of these covariates is probably not necessary in the analysis. Some efficiency will be lost for the sake of simplicity, but the sacrifice should be small.

COMPARISON OF MULTIPLE VARIABLES

If enough significance tests are done, some of the tests may be significant by chance alone. This issue of multiple comparisons includes repeated looks at the same response variable (Chapter 15) and comparisons of multiple variables. Many clinical trials have more than one response variable, and certainly several baseline variables are measured. Thus, a number of statistical comparisons are likely to be made. These would include testing for differences in entry characteristics to establish baseline comparability as well as subgroup analyses. For example, if an investigator has 100 independent comparisons, five of them, on the average, will be significantly different by chance alone if he uses 0.05 as the level of significance. The implication of this is that the investigator should be cautious in the interpretation of results if he is making multiple comparisons. The alternative is to require a more conservative significance level. As noted before, lowering the significance level will reduce the power of a trial. The issue of multiple comparisons has been discussed by Miller,[40] who reviewed many proposed approaches, and Tukey.[41]

One way to counter the problem is to increase the sample size so that a smaller significance level can be used while maintaining the power of the trial. However, in practice, most investigators could probably not afford to enroll the number of subjects required to compensate for all the possible comparisons which might be made. As an approximation, if investigators are making k comparisons, each comparison should be made at the significance level α/k. Thus, for $k = 10$ and $\alpha = 0.05$, each test would need to be significant at the 0.005 level. Sample size calculations involving a significance level of 0.005 will dramatically increase the required number of subjects. Therefore, it is more reasonable to calculate sample size based on one primary response variable comparison and be cautious in claiming significant results for other comparisons.

It is important to evaluate the consistency of the results qualitatively, and not stretch formal statistical analysis too far. Most formal comparisons should be stated in advance. Beyond that, one engages in observational data analysis to generate ideas for subsequent testing.

SUBGROUP HYPOTHESES

A common analysis technique is to subdivide or subgroup the enrolled subjects. Here the investigator looks specifically at the intervention-control comparison within one or more particular subgroups rather than the overall comparison. One of the most frequently asked questions during the design of a trial or at the time of analysis is, "Among which group of subjects is the intervention most beneficial or harmful?" It is important that subgroups be examined. Clinical trials require considerable time and effort to conduct and the resulting data deserve maximum evaluation. The hope is to refine the primary hypothesis and specify to whom, if anyone, the intervention should be recommended. Nevertheless, care must be exercised in the interpretation of subgroup findings.

As discussed earlier in this chapter, categorization of subjects by any outcome variable, eg, compliance, can lead to biased conclusions. Only baseline factors are appropriate for use in defining subgroups.

Subgroups may be identified in several ways which affect the strength of their results.[42] First, subgroup hypotheses may be specified in the study protocol. Because these are defined in advance, they have the greatest credibility. There is likely to be, however, low power for detecting differences in these subgroups. Therefore, many investigators do not pay as much attention to statistical significance as they do to the primary question. Recognizing the low chance of seeing significant differences, descriptions of subgroup effects are often qualitative. On the other hand, testing several questions can increase the chance of a Type I error. Therefore, if one were to perform tests of significance on subgroup analyses, there will be an increased probability of false positive results unless adjustments are made.

Some subgroups may be implied, but may not be explicitly stated in the protocol. For example, if randomization is stratified by age, sex, or stage of disease, it might be reasonable to infer that subgroup hypotheses related to those factors were in fact considered in advance. Of course, the same problems in interpretation apply here as with prespecified subgroups.

A third type of analysis concerns subgroups identified by other, similar trials. If one study reports that the observed difference between intervention and control appears to be concentrated in a particular subgroup of subjects, it is appropriate to see if the same findings occur in another trial, even though that subgroup was not specified in advance. Problems here include comparability of definition. It is unusual for different trials to have baseline information sufficiently similar to allow for characterization of identical subgroups.

On occasion, during the monitoring of a trial, particular subgroup findings may emerge and be of special interest. If additional subjects remain to be enrolled into the trial, one approach is to test the new subgroup hypothesis in the later subjects. With small numbers of subjects, it is unlikely that significant differences will be noted. If, however, the same pattern emerges in the newly created subgroup, the hypothesis is considerably strengthened.

The weakest type of subgrouping involves post hoc analysis,

sometimes referred to as "data-dredging" or "fishing." Such analysis is suggested by the data themselves. Because many comparisons are theoretically possible, tests of significance become difficult to interpret. Such analyses should serve primarily to generate hypotheses for evaluation in other studies.

A number of trials of beta-blocking drugs have been conducted in people with myocardial infarction. One found that the observed benefit was restricted to subjects with anterior infarctions.[43] Another claimed improvement only in subjects 65 years or younger.[44] In the Beta-Blocker Heart Attack Trial, it was observed that the greatest relative benefit of the intervention was in subjects with complications during the infarction.[45] These subgroup findings however, have not been consistently confirmed in other studies.

Regardless of how subgroups are selected, several factors can provide supporting evidence for the validity of the findings. As mentioned, similar results obtained in several studies strengthen interpretation. Internal consistency within a study is also a factor. If the same subgroup results are observed at most of the sites of a multicenter trial, they are more likely to be true. Plausible, *post hoc* biological explanations for the findings, while necessary, are not sufficient. Given almost any outcome, reasonable sounding explanations can be put forward.

Often, attention is focused on subgroups with the largest intervention-control differences. However, even with only a few subgroups, the likelihood of large but spurious differences in effects of intervention between the most extreme subgroup outcomes can be considerable.[46-49] Because large, random differences can occur, subgroup findings may easily be overinterpreted. Peto has argued that observed quantitative differences in outcome between various subgroups are to be expected, and they do not imply that the effect of intervention is truly dissimilar.[50]

It has also been suggested that, unless the main overall comparison for the trial is significant, investigators should be particularly conservative in evaluating significant subgroup findings.[7,46] Lee and colleagues conducted a simulated randomized trial, in which subjects were randomly allocated to two groups, although no intervention was initiated.[51] Despite the expected lack of overall difference, a subgroup was found which showed a significant difference.

In summary, subgroup analyses are important. However, they must be done and interpreted cautiously.

USE OF CUTPOINTS

Splitting continuous variables into two categories, for example by using an arbitrary "cutpoint," is often done in data analysis. This can be misleading, especially if the cutpoint is suggested by the data. As an example, consider the constructed data set in Table 16-5. Heart rate, in beats per minute, was measured prior to intervention in two groups of 25 subjects each. After therapies A and B were administered, the heart rate was again measured. The average changes between groups A and B are not sufficiently different from each other ($P = 0.75$) using a standard t-test. However, if these same data are analyzed by splitting the subjects into "responders" and "nonresponders," according to the magnitude of heart rate reduction, the results can be made to vary. Table 16-6 shows three such possibilities, using reductions of 7, 5, and 3 beats per minute as definitions of response. As indicated, the significance levels, using a chi-square test or Fisher's exact test,[1] change from not significant to

Table 16-5

Differences in Pre- and Post-Therapy Heart Rate, in Beats Per Minute (HR), for Groups A and B, with 25 Subjects Each

Observation Number	A			B		
	Pre HR	Post HR	Change in HR	Pre HR	Post HR	Change in HR
1	72	72	0	72	70	2
2	74	73	1	71	68	3
3	77	71	6	75	74	1
4	73	78	−5	74	71	3
5	70	66	4	71	73	−2
6	72	76	−4	73	78	−5
7	72	72	0	71	69	2
8	78	76	2	70	74	−4
9	72	80	−8	79	78	1
10	78	71	7	71	72	−1
11	76	70	6	78	79	−1
12	73	77	−4	72	75	−3
13	77	75	2	73	72	1
14	73	79	−6	72	69	3
15	76	76	0	77	74	3
16	74	76	−2	79	75	4
17	71	69	2	77	75	2
18	72	71	1	75	75	0
19	68	72	−4	71	70	1
20	78	75	3	78	74	4
21	76	76	0	75	80	−5
22	70	63	7	71	72	−1
23	76	70	6	77	77	0
24	78	73	5	79	76	3
25	73	73	0	79	79	0
Mean	73.96	73.20	.76	74.40	73.96	0.44
Standard deviation	2.88	3.96	4.24	3.18	3.38	2.66

Table 16-6
Comparison of Change in Heart Rate in Group A Versus B
by Three Choices of Cutpoints

Beats/min	<7	≥7	<5	≥5	<3	≥3
Group A	25	2	19	6	17	8
Group B	25	0	25	0	18	7
Chi-square	$P = 0.15$		$P = 0.009$		$P = 0.76$	
Fisher's exact	$P = 0.49$		$P = 0.022$		$P = 0.99$	

significant and back to not significant. This created example suggests that by manipulating the cutpoint one can observe a significance level less than 0.05 when there does not really seem to be a difference.

In an attempt to understand the mechanisms of action of an intervention, investigators frequently want to compare subjects from two groups who experience the same event. Sometimes this retrospective look can suggest factors or variables by which the subjects could be subgrouped. If some subgroup is suggested, the investigator should create that subgroup in each study group and make the appropriate comparison. For example, he may find that subjects in the intervention group who died were older than those in the control group who died. This retrospective observation might suggest that age is a factor in the usefulness of the intervention. The appropriate way to test this hypothesis would be to subgroup all subjects by age and compare intervention versus control for each age subgroup.

POOLING RESULTS ACROSS STUDIES

A controversial issue in clinical trials involves the appropriateness of pooling or combining data from more than one study. Combining data from different clinics in a multicenter trial is, in a sense, pooling. However, the same protocol is followed during the same time period at each site. Some investigators might argue that differences between clinics are so great that the issues of pooling studies apply equally to multicenter studies. Nevertheless, this section will address only pooling results from several distinct trials. Formal pooling must also be distinguished from literature surveys, which may attempt to evaluate the efficacy of particular interventions by critically reviewing published material, but which do not combine data or use statistical tests. One simple graphical method, used by May and colleagues,[52] plots the results and confidence intervals for each study in a single figure.

Pooling is usually done because no one trial may be sufficiently large or persuasive. Thus, in order to obtain the most precise estimate of efficacy, the results from a number of studies are combined. There are several recent examples of such exercises.[53-59] Pooling results from small studies may also help in planning a new clinical trial.

The actual techniques of pooling vary. One approach to pooling is to aggregate all data, ignoring the fact that different studies were involved, and produce a simple 2×2 table. While a single test statistic can be generated by this approach, information can be lost. Alternatively, the Mantel-Haenszel statistic, as described in the Appendix, could be used for dichotomous response variables. Basically, a series of 2×2 tables, one from each study, are aggregated by the Mantel-Haenszel method and a final summary statistic is obtained, adjusted for the different sources of data. This approach, or a variation of it, has been used in several pooling efforts.

Several potential problems with pooling have been identified.[60,61] One is the concern that the pooled studies differ sufficiently enough to render conclusions questionable. Obviously, study protocols, populations, interventions, types of data, and, indeed, quality, vary considerably. How much similarity is required before pooling is reasonable is a matter of debate. In rebuttal, proponents of pooling argue that as long as intervention and control are compared within each of the pooled studies, as is done in some techniques, considerable variation between studies is allowable.[56]

Another possible difficulty with pooling relates to bias. A decision process enters into the selection of trials to be pooled. Criteria on study quality or size may exclude certain studies. While these sorts of criteria are legitimate, they are often arbitrary, and may be viewed as creating a pre-determined conclusion. Also important is the fact that the set of pooled trials may be incomplete. Some trial results may not be known or even published. Those unpublished may be more heavily represented by studies showing no difference, or adverse experience, than by trials showing benefit.

Pooling, as opposed to typical literature reviews, usually puts a P-value on the conclusion. The statistical procedures may allow for calculation of a P-value, but it implies a precision which may be inappropriate. The possibility that studies may be missed and the issue of study selection may make the interpretation of the P-value tenuous. As indicated, quality of data may vary from study to study. Data from some trials may be incomplete, and perhaps not even recognized as such. Thus, only very simple and unambiguous outcome variables, such as all cause mortality, ought to be used for pooling.

Despite these problems, pooling is being seen by many as a partial solution to the extraordinary effort and cost often required to conduct an adequate individual trial. Rather than providing a solution, it perhaps ought to be viewed as a way to present existing data; a way that has strengths and weaknesses, and must be as critically evaluated as any other information. It would clearly be preferable to combine resources beforehand and collaborate in a single large study. Pooled studies cannot replace one or two well-conducted multicenter trials.

APPENDIX

Mantel-Haenszel Statistic

Suppose an investigator is comparing response rates and divides the data into a number of strata using baseline characteristics. For each stratum i, a 2×2 table is constructed.

2×2 table for i^{th} stratum

	Response		
	Yes	No	
Intervention	a_i	b_i	$a_i + b_i$
Control	c_i	d_i	$c_i + d_i$
Total	$a_i + c_i$	$b_i + d_i$	n_i

The entries a_i, b_i, c_i and d_i represent the counts in the 4 cells and n_i is the number of subjects in the ith stratum. The marginals represent totals in the various categories. The value $(a_i + c_i)/n_i$ represents the overall response rate for the ith stratum. Within the ith stratum, the rates $a_i/(a_i + b_i)$ with $c_i/(c_i + d_i)$ are compared. The standard chi-square test for 2×2 tables could be used to compare group differences in this stratum. However, the investigator is interested in "averaging" the comparison over all the strata. The method for combining several 2×2 tables over all tables or strata was described by Cochran[62] and Mantel and Haenszel.[33] The summary statistic, denoted MH, is given by

$$ MH = \frac{\{\Sigma_{i=1}^{K} [a_i - (a_i + c_i)(a_i + b_i)/n_i]\}^2}{\Sigma_{i=1}^{K} (a_i + c_i)(b_i + d_i)(a_i + b_i)(c_i + d_i)/n_i^2 (n_i - 1)} $$

The MH statistic has a chi-square distribution with one degree of freedom. Tables for this distribution are available in standard statistical textbooks. Any value for MH greater than 3.84 is significant at the 0.05 level, and any value greater than 6.63 is significant at the 0.01 level. This method is particularly appropriate for covariates that are discrete or continuous covariates that have been classified into intervals.

REFERENCES

1. Brownlee, K.A. *Statistical Theory and Methodology in Science and Engineering*. New York: John Wiley and Sons, 1965.

2. Remington, R.D., and Schork, M.A. *Statistics with Applications to the Biological and Health Sciences*. Englewood Cliffs, N.J.: Prentice-Hall, 1970.

3. Armitage, P. *Statistical Methods in Medical Research*. New York: John Wiley and Sons, 1977.

4. Hill, A.B. *Principles of Medical Statistics*, 9th Ed. New York: Oxford University Press, 1971.

5. Colton, T. *Statistics in Medicine*. Boston: Little, Brown, 1974.

6. Brown, B.W., and Hollander, M. *Statistics: A Biomedical Introduction*. New York: John Wiley and Sons, 1977.

7. Peto, R., Pike, M.C., Armitage, P. et al. Design and analysis of randomized clinical trials requiring prolonged observation of each patient. I. Introduction and design. *Br J Cancer* 34:585-612, 1976.

8. Armitage, P. The analysis of data from clinical trials. *The Statistician* 28:171-183, 1980.

9. Schwartz, D. and Lellouch, J. Explanatory and pragmatic attitudes in therapeutic trials. *J Chronic Dis.* 20:637-648, 1967.

10. Sackett, D.L., and Gent, M. Controversy in counting and attributing events in clinical trials. *N Engl J Med.* 301:1410-1412, 1979.

11. May, G.S., DeMets, D.L., Friedman, L.M. et al. The randomized clinical trial: bias in analysis. *Circulation* 64:669-673, 1981.

12. The Canadian Cooperative Study Group. A randomized trial of aspirin and sulfinpyrazone in threatened stroke. *N Engl J Med.* 299:53-59, 1978.

13. Beta-blocker Heart Attack Trial Research Group. A randomized trial of propranolol in patients with acute myocardial infarction. 1. Mortality results. *JAMA* 247:1707-1714, 1982.

14. Roberts, R., Croft, C., Gold, H.K., et al. Effect of propranolol on myocardial infarct size in a randomized blinded multicenter trial. *N Engl J Med.* 311:218-225, 1984.

15. Collaborative Group on Antenatal Steroid Therapy. Effect of antenatal dexamethasone administration on the prevention of respiratory distress syndrome. *Am J Obstet Gynecol.* 141:276-287, 1981.

16. The Anturane Reinfarction Trial Research Group. Sulfinpyrazone in the prevention of sudden death after myocardial infarction. *N Engl J Med.* 302:250-256, 1980.

17. Temple, R. and Pledger, G.W. The FDA's critique of the Anturane Reinfarction Trial. *N Engl J Med.* 303:1488-1492, 1980.

18. Anturane Reinfarction Trial Policy Committee. The Anturane Reinfarction Trial: reevaluation of outcome. *N Engl J Med.* 306:1005-1008, 1982.

19. Coronary Drug Project Research Group. Influence of adherence to treatment and response of cholesterol on mortality in the Coronary Drug Project. *N Engl J Med.* 303:1038-1041, 1980.

20. Verter, J. and Friedman, L. Adherence measures in the Aspirin Myocardial Infarction Study (AMIS). (abstract) *Controlled Clin Trials* 5:306, 1984.

21. Pizzo, P.A., Robichaud, K.J., Edwards, B.K., et al. Oral antibiotic prophylaxis in patients with cancer: a double-blind randomized placebo-controlled trial. *J. Pediatrics* 102:125-133, 1983.

22. Canner, P.L. Monitoring clinical trial data for evidence of adverse or beneficial treatment effects. Edited by J.P. Boissel and C.R. Klimt. In

264

Multicenter Controlled Trials: Principles and Problems. Paris: INSERM, 1979.

23. Wilcox, R.G., Roland, J.M., Banks, D.C., et al. Randomised trial comparing propranolol with atenolol in immediate treatment of suspected myocardial infarction. *Br Med J.* 1:885–888, 1980.

24. Detre, K. and Peduzzi, P. The problems of attributing deaths of nonadherers: the VA coronary bypass experience. *Controlled Clin Trials* 3:355–364, 1982.

25. Coronary Drug Project Research Group. Clofibrate and niacin in coronary heart disease. *JAMA* 231:360–381, 1975.

26. Lipid Research Clinics Program. The Lipid Research Clinics Coronary Primary Prevention Trial results. 1. Reduction in incidence of coronary heart disease. *JAMA* 251:351–364, 1984.

27. Kruskal, W.H. Some remarks on wild observations. *Technometrics* 2:1–3, 1960.

28. Dixon, W.J. Processing data for outliers. *Biometrics* 9:74–89, 1953.

29. Dixon, W.J. Rejection of observations. Edited by A.E. Sarhan and B.G. Greenberg. In *Contributions to Order Statistics*. New York: John Wiley and Sons, 1962.

30. Grubbs, F.E. Procedures for detecting outlying observations in samples *Technometrics* 11:1–21, 1969.

31. Canner, P.L., Huang, Y.B., and Meinert, C.L. On the detection of outlier clinics in medical and surgical trials: I. Practical considerations. *Controlled Clin Trials* 2:231–240, 1981.

32. Canner, P.L., Huang, Y.B., and Meinert, C.L. On the detection of outlier clinics in medical and surgical trials: II. Theoretical considerations. *Controlled Clin Trials* 2:241–252, 1981.

33. Mantel, N., and Haenszel, W. Statistical aspects of the analysis of data from retrospective studies of disease. *J Natl Cancer Inst.* 22:719–748, 1959.

34. Bishop, Y.M., Fienberg, S.E., and Holland, P.W. *Discrete Multivariate Analysis: Theory and Practice.* Cambridge: MIT Press, 1975.

35. Mitchell, R.J. What constitutes evidence on the dietary prevention of coronary heart disease? Cosy beliefs or harsh facts? *Int J Cardiol.* 5:287–298, 1984.

36. Report from the Committee of Principal Investigators. A co-operative trial in the primary prevention of ischaemic heart disease using clofibrate. *Br Heart J.* 40:1069–1118, 1978.

37. Weiss, G.B., Bunce, H. III, and Hokanson, J.A. Comparing survival of responders and nonresponders after treatment: a potential source of confusion in interpreting cancer clinical trials. *Controlled Clin Trials* 4:43–52, 1983.

38. Anderson, J.R., Cain, K.C., and Gelber, R.D. Analysis of survival by tumor response. *J Clin Oncology* 1:710–719, 1983.

39. Cox, D.R. Regression models and lifetables. *J R Stat Soc.* Series B 34:187–202, 1972.

40. Miller, R.G., Jr. *Simultaneous Statistical Inference.* New York: McGraw-Hill, 1966.

41. Tukey, J.W. Some thoughts on clinical trials, especially problems of multiplicity. *Science* 198:679–684, 1977.

42. Ingelfinger, J.A., Mosteller, F., Thibodeau, L.A., and Ware, J.H. *Biostatistics in Clinical Medicine.* New York: MacMillan, 1983, 234–235.

43. Multicentre International Study. Improvement in prognosis of myocardial infarction by long-term beta-adrenoreceptor blockade using practolol. *Br Med J* 3:735–740, 1975.

44. Andersen, M.P., Bechsgaard, P., Frederiksen, J., et al. Effect of alprenolol on mortality among patients with definite or suspected acute myocardial infarction. *Lancet* ii:865–868, 1979.

45. Furberg, C.D., Hawkins, C.M., Lichstein, E. Effect of propranolol in postinfarction patients with mechanical or electrical complications. *Circulation* 69:761–765, 1984.

46. Simon, R. Patient subsets and variation in therapeutic efficacy. *Br J Clin Pharmac* 14:473–482, 1982.

47. Moses, L.E. The series of consecutive cases as a device for assessing outcomes of intervention. *N Engl J Med* 311:705–710, 1984.

48. Ingelfinger, J.A., Mosteller, F., Thibodeau, L.A., and Ware, J.H. *Biostatistics in Clinical Medicine.* New York: MacMillan, 1983, 255–258.

49. Furberg, C.D., and Byington, R.P. What do subgroup analyses reveal about differential response to beta-blocker therapy? The Beta-Blocker Heart Attack Trial Experience. *Circulation* 67(Suppl I):I-98–I-101, 1983.

50. Peto, R. Statistical aspects of cancer trials. Edited by K.E. Halnan. In *Treatment of Cancer.* London: Chapman and Hall, 1982.

51. Lee, K.L., McNeer, J.F., Starmer, C.F. et al. Clinical judgment and statistics: lessons from a simulated randomized trial in coronary artery disease. *Circulation* 61:508–515, 1980.

52. May, G.S., Furberg, C.D., Eberlein, K.A., and Geraci, B.J. Secondary prevention after myocardial infarction: a review of short-term acute phase trials. *Prog Cardiovasc Dis.* 25:335–359, 1983.

53. Chalmers, T.C., Matta, R.J., Smith, H. Jr., and Kunzler, A.M. Evidence favoring the use of anticoagulants in the hospital phase of acute myocardial infarction. *N Engl J Med.* 297:1091–1096, 1977.

54. Baum, M.L., Anish, D.S., Chalmers, T.C., et al. A survey of clinical trials of antibiotic prophylaxis in colon surgery: evidence against further use of no-treatment controls. *N Engl J Med.* 305:795–799, 1981.

55. DeSilva, R.A., Hennekens, C.H., Lown, B., and Casscells, W. Lignocaine prophylaxis in acute myocardial infarction: an evaluation of randomised trials. *Lancet* ii:855–858, 1981.

56. Stampfer, M.J., Goldhaber, S.Z., Yusuf, S., et al. Effect of intravenous streptokinase on acute myocardial infarction. Pooled results from randomized trials. *N Engl J Med.* 307:1180–1182, 1982.

57. Budetti, P.P. and McManus, P. Assessing the effectiveness of neonatal intensive care. *Medical Care* 20:1027–1039, 1982.

58. Canner, P.L. Aspirin in coronary heart disease: comparison of six clinical trials. *Isr J Med Sci.* 19:413–423, 1983.

59. Messer, J., Reitman, D., Sacks, H.S., et al. Association of adrenocorticosteroid therapy and peptic ulcer-disease. *N Engl J Med.* 309:21–24, 1983.

60. Elashoff, J.D. Combining results of clinical trials (editorial). *Gastroenterology* 75:1170–1174, 1978.

61. Goldman, L., and Feinstein, A.R. Anticoagulants and myocardial infarction: the problems of pooling, drowning, and floating. *Ann Intern Med.* 90: 92–94, 1979.

62. Cochran, W.G. Some methods for strengthening the common chi-square tests. *Biometrics* 10:417–451, 1954.

CHAPTER 17

Closeout

In any clinical trial, appropriate plans for the closeout phase need to be developed. This phase starts with the final follow-up visit of the first subject enrolled and lasts until all analyses have been completed. It is evident that well before the scheduled end of the trial, there needs to be a fairly detailed plan for this phase if the study is to terminate in an orderly manner. One must be prepared to modify this plan since unexpected results, either beneficial or harmful, may require that the trial be stopped early. Bell et al contrast the closeout procedures and problems in a study requiring early termination with those from a study stopping on schedule.[1] Krol describes the closeout of the Coronary Drug Project, a large-scale, multi-armed trial.[2] It is possible that only certain subgroups of subjects will be taken off intervention earlier than scheduled.[3-5] This chapter will address a number of closeout topics. Although many of them relate to large single-center or multicenter trials, they may be applicable also to smaller studies. Topics discussed include technical procedures for the termination of the trial, post-study follow-up, dissemination of trial results, and clean-up and storage of data. Obviously, the details of the closeout plan have to be tailored to the particular trial.

FUNDAMENTAL POINT

The closeout of a clinical trial is usually a fairly complex process that requires careful planning if it is to be accomplished in an orderly fashion.

TERMINATION PROCEDURES

If each subject in a clinical trial is to be followed for a fixed period of time, the closeout phase will be of at least the same duration as the enrollment phase. In many cases, this means several months to years. Terminating the follow-up of some study subjects while others are still being actively followed can create problems in long-term trials, and this closeout design may not be desirable. An alternative, and frequently used, plan involves following all subjects to a common termination date, or when this is not feasible, to a compressed closeout period. The termination date is determined by two factors; the date that the last subject is enrolled, and the minimum length of follow-up period that the protocol calls for. Therefore, if the last subject is enrolled March 15, 1983, and the minimum scheduled follow-up is three years, the common termination date would be March 15, 1986.

There are several advantages to this approach. First, in most blinded trials, the code for each subject is broken at the last scheduled follow-up visit. If the unblinding occurs over a span of several months or years, as in the former strategy, there is the possibility of accidentally breaking the blind of subjects still actively followed in the trial. For example, more than one subject in a drug trial may be given study medication with the same bottle code (Chapter 6). Breaking the blind for one subject of four who get bottles with the same code will unblind the other three. In addition, the investigator may start associating a certain symptom or constellation of symptoms and signs with particular drug codes.

Second, the investigator's obligation to a subject means that at the final follow-up visit the investigator needs to advise the subject regarding possible continued intervention. If the closeout is extended over a long period, as it would be if each subject were followed for the same duration, any early recommendation to an individual subject would have to be based on incomplete follow-up data which may not reflect the trial conclusion. Moreover information could "leak" to subjects still in the trial, thus affecting the integrity of the trial. Although it is highly desirable to provide each subject with a recommendation regarding continued treatment, doing so may not be possible until the study is completely over. When unblinding occurs over a span of months or years, this leaves the investigator in the uncomfortable position of ending a subject's participation in the trial and asking him to wait months before he can be told the study results and advised what to do. If the incomplete results are clearcut, it can be easy to arrive at such recommendations. However, in such an instance, the investigator would be confronted with an ethical dilemma. How can he recommend that a subject start, continue, or discontinue a new intervention while keeping other subjects in the trial?

Third, adopting the common termination design adds to the power of the trial by extending the follow-up period beyond the minimum time for all but the last subject enrolled. In a trial with two years of recruitment, the additional follow-up period would increase by an average of one year, assuming a uniform recruitment rate. In addition, in terms of subject-years of follow-up, this approach might be more cost-efficient when clinic staff is supported solely by the sponsor of the trial. With all subjects followed to the end of the study, full support of personnel can be justified until all subjects have been seen for the last time. In trials where the subjects are phased out after a fixed time of follow-up, an increase in the staff/subject ratio may be unavoidable.

Despite the problems with following all subjects for a fixed length of time, this may be the preferable approach in certain trials, particularly those with a relatively short follow-up phase. In such studies, there is no realistic alternative. Seeing subjects for more visits than are scientifically necessary is hard to justify and will obviously add to trial cost and time involvement. In addition, a limitation of the common termination approach is that it may not be feasible to conduct a large number of closeout visits in a short time. Depending on the content of the last visit and availability of staff and weekly clinic hours, seeing 100 to 150 subjects at a clinic may require several weeks. A decision on the type of follow-up plan should be based on the scientific question as well as logistics.

The last follow-up visit is the most important visit after enrollment. It is particularly so in trials where the main response variables are continuous variables such as laboratory or electrocardiographic data. By necessity, the response variable data must be obtained for each subject at the last follow-up visit because it marks the end of follow-up. If the subject fails to show up for the last visit, the investigator will lack data. When the response variable is the occurrence of a specific event, such as a fatal myocardial infarction, the situation is somewhat different. The occurrences of such response variables are usually recorded throughout the trial. Also, the information can be obtained, in certain instances, without having the subject complete a visit.

If a subject suffers an event after his last follow-up visit but before all subjects have been seen for the final visit, the investigator must decide whether or not that response variable should be included in the data analysis. The simplest solution is to let the last follow-up visit denote each subject's termination of the trial. For subjects who do not show up for the last visit, the investigator has to decide when to make the final ascertainment. If death is the primary response variable, vital status is usually determined as of the last day of the follow-up phase. Although this may introduce a slight bias, the last day is used because the investigator does not know with certainty whether a subject is going to miss the last visit until the last possible date for this visit has passed.

It is important in any trial to obtain, to the extent possible, response variable data on every enrolled subject. The uncertainty of the results rises as the number of subjects for whom response variable data are missing increases. For example, assume that death from any cause is the primary response variable in a trial and the observed mortality is 15% in one group and 10% in the other group. Depending on study size, this group difference might be statistically significant. However, if 10% of the subjects in each group were lost to follow-up, the observed outcome of the trial can be questioned. It cannot be assumed that the mortality experience among those lost to follow-up is the same as for those who stayed in the trial, or that those lost to follow-up in one group have a mortality experience identical to those lost to follow-up in the other group. Equally important, there should be no differential assessment in the study groups. Therefore, every effort should be made to ensure as complete as possible final ascertainment of response variables.

A number of means have been used to track subjects and determine their vital status. These include the use of a person's social security number, relatives, or employers. In countries with national death registries, which now include the United States, mortality surveillance is simpler and probably more complete than in countries without such registries. Agencies that specialize in locating people have been used in several trials. This is a very sensitive area, since a search can be looked upon as an intrusion into the privacy of the subject. The integrity of a trial and its results plus the subject's initial agreement to participate in the trial have to be weighed against a person's right to protect his privacy. Investigators may want to include in the informed consent form a sentence stating that the participant agrees to have his vital status determined at the end of the trial even if he has by then stopped participating actively.

POST-STUDY FOLLOW-UP

Subject follow-up after the termination of a trial is comprised of two activities. One involves short-term follow-up during a "cooling-off" period when subjects are off intervention but still being followed. The other is long-term follow-up monitoring of possible toxicity or benefit. These activities are separate from the moral obligation of the investigator to facilitate, when necessary, a subject's return to the usual medical care system, to ensure that study recommendations are communicated to his private physician, and at times to continue the subject on a beneficial new intervention. A cooling-off period to find out how soon laboratory values or symptomatology return to pre-trial level or status ought to be considered in trials where the intervention is being stopped at the last follow-up visit. A cooling-off period is particularly important when the trial results are negative or inconclusive, when the intervention is not available for non-study use or when no recommendations as to continued intervention can be made. The effect of the intervention may last long after a drug has been stopped, and side effects or "toxic" changes revealed by laboratory measurements may not disappear until weeks after intervention has ended. For certain drugs, such as beta-blockers, the intervention should not be stopped abruptly. A tapering of the dosage may require additional clinic visits. After dextrothyroxine (one of the interventions in the Coronary Drug Project) was discontinued, cases of clinical hypothyroidism requiring therapy were noted.[6]

Post-study follow-up of subjects is a rather complex process in most countries. First, the investigator has to decide what should be monitored. Mortality surveillance can be cumbersome and is worth undertaking only if there is a reasonable expectation of getting an almost complete record of vital status. Usually, the justification for long-term post-study surveillance is based on a trend or unexpected finding in the trial or from a finding from another source.

Obtaining information on non-fatal events is usually more complicated and, in general, its value is questionable. However, an illustration that post-study follow-up for toxicity can prove valuable is the finding of severe toxic effects attributed to diethylstilbestrol (DES). The purported carcinogenic effect occurred 15 to 20 years after the drug was administered and occurred in female offspring who were exposed in utero.[7]

In 1975, a trial of clofibrate in people with elevated lipids[8] reported an excess of cases of cancer in the clofibrate group compared to the control group. The question was raised whether the subjects assigned to clofibrate in the Coronary Drug Project[7] also showed an increase in the cancer incidence. This was not the case. However, since the number of cancer deaths during the Coronary Drug Project was small and the follow-up was relatively short (five years), the possibility of a post-trial cancer mortality surveillance was considered. Because only 3% of the deaths during the study were cancer-related, this survey was not thought to be feasible. Subsequently, the WHO study of clofibrate reported that all cause mortality was increased in the intervention group.[10] At that time, the Coronary Drug Project investigators decided that post-study follow-up was scientifically and ethically important, and such a study was undertaken. The example brings up a question: Should investigators of large-scale clinical trials make arrangements for surveillance in case, at some future time, the need for such a study were to arise? The implementation of any post-study surveillance plan raises complications. A key one is to find a way of keeping subjects' names and addresses in a central registry without infringement upon the privacy of the individuals. The investigator must also decide, with little evidence, on the optimal duration of surveillance after the termination of a trial (eg, 2, 5 or 20 years).

A second issue of post-study surveillance is related to a possible beneficial effect of intervention. In any intervention trial, assumptions must be made with respect to time between initiation of intervention and the occurrence of full beneficial effect. For many drugs, this so-called "lag-time" is assumed to be zero. However, if the intervention is a lipid-lowering drug or a dietary change and the response variable is coronary mortality, the lag-time might be several years. The problem in contemplating a trial of such an intervention is that the maximum practical follow-up may not be long enough for a beneficial effect to appear. Extended surveillance after completion of the follow-up phase may be considered in such studies. In fact, the Coronary Drug Project follow-up referred to above showed unexpected benefit in one of the intervention groups. At the conclusion of the trial, the nicotinic acid subjects had had significantly fewer nonfatal reinfarctions, but no difference in survival was detected.[9] Total mortality, after an average 6.5 years in the trial on drug, plus an additional 5 years after the trial, however, was significantly less in the group assigned to nicotinic acid than in the placebo group.[11]

There are several interpretations of this finding. It is possible that this observation is real, and that the benefit simply took longer than expected to appear. Also, the earlier reduction in nonfatal myocardial infarction may have finally affected prognosis. Of course, the results may be due to chance. A major difficulty in interpreting the data relates to the lack of knowledge about what the subjects in the intervention and control groups did with respect to lipid lowering and other regimens in the intervening 5 years. Although there was no reason to expect that there was differential use of any intervention affecting mortality, such could have been the case.

As already pointed out, knowledge of the response variable of interest for almost every subject is required if long-term surveillance after completion of regular follow-up is to be worthwhile. The degree of completeness attainable depends on several factors, such as the length of surveillance time, the community where the trial was conducted and the aggressiveness of the investigator.

DATA CLEAN-UP AND VERIFICATION

Despite attempts to collect complete, consistent, and error-free data, perfection is unlikely to be achieved. Any monitoring system will reveal missing forms, unanswered items on forms and conflicting data.

The importance of continuous monitoring of study forms and data throughout a trial is pointed out elsewhere (Chapter 10). It helps to minimize the job of cleaning up data at the end of the study. Nevertheless, some final data editing will be necessary. The timing of this process is important. Data editing should be initiated as soon as possible, because it is difficult to get full staff cooperation after a trial and its funding are over. This may be especially difficult in a multicenter trial, where each investigator tends to pursue other interests once the study ends. It is also necessary to be realistic in the clean-up process. This means "freezing" the files at a reasonable time after the termination of subject follow-up and accepting some incomplete data. Obviously, the efforts during clean-up should be directed toward the most important areas; those crucial to answering the primary question.

Any clinical trial may be faced with having its results reviewed, questioned and even audited. Traditionally, this review has been a scientific one. However, since special interest groups may want to look at the data, the key results should be properly verified, documented and filed in an easily retrievable manner. The extent of this additional documentation of important data will depend on the design of each trial. Various models have been used. In one large multicenter trial,[12] a second Data Coordinating Center was established. Duplicates of the key study forms were submitted to this center, which generated separate data reports. This is obviously a costly approach. In the Anturane Reinfarction Trial,[13] an outside group of experts audited the data before the results were published. In another trial, the regular procedure for reporting the primary response variable (death) was to complete and submit a Death Notification Form to the Data Coordinating Center. To document the primary response variable, each investigator was asked at the end of follow-up to send a list of all deceased subjects along with date of death to an office independent of the Coordinating Center. The important feature here is that the verification was handled by an independent group.

Verification of data may be time-consuming; thus it can conflict with the desire of the investigator to publish his findings as early as possible. While publication of important information should not be delayed unnecessarily, results should not be put into print before key data have been verified.

STORAGE OF STUDY MATERIAL

There are three principal reasons for storage of study material after a trial has been terminated. The first two, dealing with post-study harm or benefit and with the possibility of an audit, have been mentioned. The third reason is scientific. In planning for a new trial, an investigator may want to obtain unpublished data from other investigators who have conducted trials in a similar population or tested the same intervention. Similarly, in preparation of a review article or a paper on the natural history of a disease, an investigator may want to obtain additional information from published trials. Tables and figures in scientific papers seldom include everything that may be of interest. No uniform mechanism exists today for getting access to such study material from terminated trials. If information is obtained, it may not be in a reasonable and easily retrievable form. Substantial cooperation is usually required from the investigators originally involved in the data collection and analysis.

The responsibility of an investigator is to ensure that relevant data are made accessible in an organized fashion after the termination of a trial. That is not to say data need be made accessible while a study is in progress; only after the data analyses have been completed and the results published. The risk is that other investigators might analyze the data and arrive at different interpretations of the results. However, further analysis and discussion of different interpretations of trial data is usually scientifically sound and ought to be encouraged.

In most trials, an excess of study material is collected and it may not make sense to store everything. The investigator has to consider logistics, the length of the storage period and cost. He also has to keep in mind that biological material, for example, deteriorates with time, data tapes require regular maintenance and laboratory methodologies change, making future comparisons of data difficult.

One set of documents (eg, trial protocol, manual of procedures, study forms) and the analytic material (eg, data tapes) should be kept by the investigator. In addition, a list containing identifying information for all subjects who participated in a trial ought to be stored at the institution where the investigation took place. Local regulations sometimes require that individual subject data (copies of study forms, laboratory reports, electrocardiograms, and x-rays) be filed for a defined period of time with the subject's medical records. Storage of these data on microfilm may ease the problem of inadequate space. The actual trial results and their interpretation are usually published and can be retrieved through a library search. Exceptions are obviously findings which never reach the scientific literature. It would be desirable in these cases if draft manuscripts along with other documentation and analytic material were filed by the investigator.

DISSEMINATION OF RESULTS

The reporting of findings from a small single-center trial is usually straightforward. The individual subjects are often told about the results at the last follow-up visit or shortly afterward, and the medical community is informed through scientific publications. However, there are situations that make the dissemination of findings difficult, especially the order in which the various interested parties are informed. In studies where the subjects are referred by physicians not involved in the trial, the investigator has an obligation to tell these physicians about the conclusions. In multicenter trials with clinics geographically scattered, all investigators have to be brought together or learn the results. In certain instances, the sponsoring party has a desire to make the findings known publicly at a press conference. However, although an early press conference followed by an article in a newspaper may be politically important to the sponsor of the trial, it could offend the subjects, the referring physicians and the medical community. They may all feel that they have a right to be informed before the results are reported in the lay press.

One sequence has been suggested by Klimt and Canner.[6] First, the study leadership informs the other investigators who, in turn, inform the study subjects. The private physicians of the subjects are also told, in confidence, of the findings. The results are then published in the scientific press, after which they may be more widely disseminated in other forums.

REFERENCES

1. Bell, R.L., Curb, J.D., Friedman, L.M., et al. Termination of clinical trials: The Beta-Blocker Heart Attack Trial and the Hypertension Detection and Follow-up Program experience. *Controlled Clin Trials* 6:102–111, 1985.

2. Krol, W.F. Closing down the study. *Controlled Clin Trials* 4:505–512, 1983.

3. Coronary Drug Project Research Group. The Coronary Drug Project: findings leading to further modifications of its protocol with respect to dextrothyroxine. *JAMA* 220:996–1008, 1972.

4. Coronary Drug Project Research Group. The Coronary Drug Project: findings leading to discontinuation of the 2.5-mg/day estrogen group. *JAMA* 226:652–657, 1973.

5. Diabetic Retinopathy Study Research Group. Preliminary report on effects of photocoagulation therapy. *Am J Ophthalmol.* 81:383–396, 1976.

6. Klimt, C.R., and Canner, P.L. Terminating a long-term clinical trial. *Clin Pharmacol Ther.* 25:641–646, 1979.

7. Herbst, A.L., Ulfelder, H., and Poskanzer, D.C. Adenocarcinoma of the vagina: association of maternal stilbestrol therapy with tumor appearance in young women. *N Engl J Med.* 284:878–881, 1971.

8. Report from the Committee of Principal Investigators. A co-operative trial in the primary prevention of ischaemic heart disease using clofibrate. *Br Heart J.* 40:1069–1118, 1978.

9. Coronary Drug Project Research Group. Clofibrate and niacin in coronary heart disease. *JAMA* 231:360–381, 1975.

10. Committee of Principal Investigators. WHO cooperative trial on primary prevention of ischaemic heart disease using clofibrate to lower serum cholesterol: mortality follow-up. *Lancet* ii:379–385, 1980.

11. Canner, P.L. Mortality in Coronary Drug Project patients during a nine-year post-treatment period. (abstract) *JACC* 5:442, 1985.

12. Persantine-Aspirin Reinfarction Study Research Group. Persantine and aspirin in coronary heart disease. *Circulation* 62:449–461, 1980.

13. Anturane Reinfarction Trial Research Group. Sulfinpyrazone in the prevention of cardiac death after myocardial infarction. *N Engl J Med*. 298: 289–295, 1978.

CHAPTER 18

Reporting and Interpretation of Results

The final phase in any experiment is to interpret and report the results. Finding the true answer to a challenging question is the ultimate goal of any research endeavor. To achieve this, the investigator has to review his results critically and avoid the temptation of overinterpreting them. He is in the privileged position of knowing the quality and limitations of the data better than anyone else. He, therefore, has the responsibility for presenting them clearly and concisely, together with any issues which might bear on their interpretation.

A study may be reported in a scientific journal, but this is in no way an endorsement of its results or conclusions. Even if the journal uses referees to assess each prospective publication, there is no assurance that they have sufficient experience and knowledge of the design, conduct and analysis of the reported study.[1] As pointed out by the Editor of the New England Journal of Medicine,[2] "In choosing manuscripts for publication we make every effort to winnow out those that are clearly unsound, but we cannot promise that those we do publish are absolutely true. . . . Good journals try to facilitate this process [of medical progress] by identifying noteworthy contributions from among the great mass of material that now overloads our scientific communication system. Everyone should understand, however, that this evaluative function is not quite the same thing as endorsement." In the end, it is up to the reader of a scientific article to critically assess it and to decide how to best make use of the reported findings.

The issues discussed in this chapter need to be considered both by the investigator when preparing a scientific paper and by readers when evaluating it. Other guidelines on how to appraise a clinical trial report also exist.[3-5]

FUNDAMENTAL POINT

The investigator has an obligation to review critically his study and its findings and to present sufficient information so that readers can properly evaluate the trial.

Clinical trials are similar to other experiments when it comes to reporting results. The text should be explicit, and care must be taken in preparing and designing tables and figures. The primary and any secondary questions or hypotheses need to be clearly stated, as well as the rationale for the selection of the response variables and how they were determined and verified. The sources of study subjects ought to be indicated, along with a detailed description of entry and exclusion criteria. Information on the number of subjects screened and the proportion of those who were enrolled is helpful in evaluating to what extent the findings can be generalized. The nature of the intervention and of the follow-up schedule needs to be provided. Other key design features should be specified, including the method for group assignment (eg, randomization) and type of blindness. A brief review of the sample size calculation, and the assumptions that went into this calculation is also useful. Furthermore, specific questions need to be answered. These are: (a) Did the trial work as planned? (b) How do the findings compare with those from other studies? (c) What is the clinical impact of the findings? Reporting standards for clinical trials have been suggested by Mosteller et al.[6]

DID THE TRIAL WORK AS PLANNED?

The foundation of any clinical trial is the concept that the study groups were comparable initially, and that differences between the groups over time reasonably can be attributed to the effect of intervention. Although randomization is the preferred method used to obtain baseline comparability, it does not guarantee balance at baseline in the distribution of known or unknown prognostic factors. Baseline imbalance is fairly common in small trials but may also exist in large trials. Therefore, documentation and evaluation of baseline comparability are essential. Should the trial be non-randomized, the credibility of the findings hinges even more upon an adequate documentation of this comparability. For each group, baseline data should include means and standard deviations of known and possible prognostic factors. Note that the absence of a statistically significant difference for any of these factors does not mean that the groups are balanced. In small studies large differences are required in order to reach statistical significance. In addition, small trends for individual factors can have an impact if they are in the same direction. A multivariate analysis may be advantageous in evaluating balance. Of course, the fact that major prognostic factors may be unknown will produce some uncertainty with regard to baseline balance. Adjustment of the findings on the basis of observed baseline imbalance should be performed and any difference between unadjusted and adjusted analyses should be carefully explained.

Double-blindness is a desirable feature of a clinical trial design because, as already discussed, it diminishes bias in the reporting and assessment of response variables. However, few studies are truly double-blind to all parties from start to finish. While an individual side-effect may be insufficient to unblind the investigator, a constellation of effects often reveals the group assignment. A specific drug effect (eg, a marked fall in blood pressure in an antihypertensive drug trial) — or the absence of such an effect — might also indicate which is the treatment group. Although the success of blinding may be difficult for the investigator to assess, the evaluation must be done. Readers of a publication ought to be informed about the degree of unblinding. An evaluation such as that provided by Karlowski and colleagues[7] is commendable.

In estimating sample size, assumptions may be made regarding the rate of noncompliance. Throughout follow-up, efforts are made to maintain optimal compliance with the intervention under study and to monitor compliance. When interpreting the findings, one can then gauge whether the initial assumptions were borne out by what actually happened. When assumptions have been too optimistic, the ability of the trial to test adequately the primary question may be less than planned. The study results must be reported and discussed with the power of the trial in mind. Noncompliance is usually a minor concern in trials showing a beneficial effect of a specific intervention. Of course, it may be argued that the intervention would have been even more beneficial had compliance been higher. On the other hand, if all subjects (including those who for various reasons did not comply entirely with the dosage schedule of a trial) had been on full dose, there could have been further adverse or toxic effects in the intervention group.

Also of interest is the comparability of groups during the follow-up period with respect to concomitant intervention. Use of drugs other than the study intervention, changes in lifestyle and general medical care — if they affect the response variable — need to be measured. Of course, as mentioned in Chapter 16, adjustment on post-randomization variables is inappropriate. As a consequence, when imbalances exist, the study results must be interpreted cautiously.

Was the hypothesis adequately tested? This question should be asked when the results of a trial indicate no statistically significant difference between the study groups. A trial of an intervention with a true beneficial effect may give inconclusive results for several reasons. The dose of the studied intervention may have been too low or too high; the technical skills of those providing the intervention (eg, surgical procedure) may have been inadequate; the sample size may have been too small, giving the trial insufficient power to test the hypothesis (Chapter 7); there may have been major compliance problems; or concomitant intervention may have reduced the effect that would otherwise have been seen. The authors and the readers of a paper must consider all these points before accepting the conclusion of no difference between groups.

Are there any limitations of the findings? One needs to know the degree of completeness of data in order to evaluate a trial. A typical shortcoming, particularly in long-term trials, is that the investigator may lose track of some subjects. These subjects are usually different from those who remain in the trial, and their event rate may not be the same. Vigorous attempts should be made to keep the number of persons lost to follow-up to a minimum. The credibility of the findings may be questioned in trials in which the number of subjects lost to follow-up is large in relation to the number of events. A conservative approach in this context is to assume the "worst case." The "worst case" approach assumes the occurrence of an event in each subject lost to follow-up in the group with lower incidence of the response variable, and it assumes no events in the comparison group. After application of the "worst case" approach, if the overall conclusions of the trial remain unchanged, they are strengthened. However, if the results could be altered to the extent of changing the conclusions, the trial may have less credibility. The degree of confidence in the conclusion will depend upon the extent to which the outcome could be altered by the missing information.

As addressed in Chapter 16, results may be questioned if subjects randomized into a trial are withdrawn from the trial analysis. Different schools of thought exist about the validity of removing subjects subsequent to randomization. What would be desirable, however, is for investigators who support the concept of allowing withdrawals from the analysis to include in their reporting of results analyses both with, and without, withdrawals. If both analyses give approximately the same result, the findings are confirmed. However, if the results of the two analyses differ, the interpretation of the study may remain unclear, and reasons for the differences need to be explored.

In evaluating possible benefit of an intervention, more than one response variable is often assessed which raises the issue of multiple comparisons (Chapter 16). In essence, the chance of finding a nominally statistically significant result increases with the number of comparisons. This is true whether there are repeated comparisons for the same response variable or if combinations of various response variables are tested. The potential impact of this on the findings and conclusion of a trial ought to be considered. A conservative approach in the interpretation of statistical tests is again recommended. When several comparisons have been made, a more extreme statistic might be required before a statistically significant difference could be claimed. Readers of a report should be informed of the total number of comparisons made during a trial and in the analysis phase. Further caution is required in the interpretation of results based on continuous response variables which are treated as dichotomous variables.

The main objective of any trial is to answer the primary question. Findings related to one of the secondary questions may be interesting, but they should be put in the proper perspective. Are the findings for the related primary and secondary response variables consistent? If not, attempts ought to be made to explain discrepancies. Explaining inconsistencies was particularly important in the Cooperative Trial in the Primary Prevention of Ischaemic Heart Disease.[8] In that trial, the intervention group showed a statistically significant reduction in the incidence of major ischemic heart disease (primary response variable) but a significant increase in mortality for any cause (secondary response variable). In all studies, evidence for possible serious adverse effects from the intervention needs to be presented. In the final conclusion, the overall benefit should be weighed against the risk of harm. This, however, is infrequently done (Chapter 11).

Whatever the reported effect of an intervention, be it statistically significant or non-significant, there must always be some uncertainty, since a trial involves only a sample of the study population. Uncertainty can be expressed by presenting the observed difference between groups along with the standard deviation of the difference. Confidence limits around the observed difference might also be presented.[1] Approximately 95% confidence limits are typically constructed by adding and subtracting two standard deviations from the observed difference. Thus, there is 95% confidence that the true difference between the groups lies within two standard deviations on either side of the observed difference.

HOW DO THE FINDINGS COMPARE WITH RESULTS FROM OTHER STUDIES?

The findings from a clinical trial should be placed in the context of current knowledge. Are they consistent with knowledge of basic science, including presumed mechanism of action of the intervention? Proof of why an intervention works is seldom achieved. Nevertheless, when the outcome can be explained in terms of known biological actions, the conclusions are strengthened. Do the findings confirm the results of studies with similar interventions or different interventions in similar populations? Generally, credibility increases with the proportion of good independent studies that come to the same conclusion. Inconsistent results are not uncommon in research. In such cases, the problem for both the investigator and the reader is to try to determine the true effect of an intervention. How and why results differ need to be explored. Confidence limits provided in the report of a trial are useful to the reader when he is faced with having to compare the results from more than one trial. He can then more readily assess whether the results of different trials could, in fact, be consistent.

WHAT IS THE CLINICAL IMPACT
OF THE FINDINGS?

The findings of any clinical trial are, in a very strict sense, applicable only to the subjects who participated in that trial. It is appropriate, of course, to generalize the results to the study population, that is, those people meeting the eligibility criteria. The next step, suggesting that the trial results be applied to a more general population (the majority of which would not even meet the eligibility criteria of the trial) is more tenuous. The reader must judge for himself whether or not such an extrapolation is appropriate.

A similar argument applies to the intervention itself. How general are the findings? If the intervention involved a special procedure, such as surgery or counseling, is its application outside the trial setting likely to produce the same response? In a drug trial the question of dose-effect relationship is often raised. Would a higher dose of the drug have given different results? Can the same claims be made for different drugs that have a similar structure or pharmacological action? Can the results of an intervention be generalized even more broadly? For example, are all lipid-lowering regimens (diet, drug, or partial ileal bypass surgery) equivalent in testing the general hypothesis that lowering of serum cholesterol reduces the incidence of coronary heart disease, or is the answer specific to the mode of intervention? For a further discussion of generalization, see Chapter 3.

As with all research, a clinical trial will often raise as many questions as it answers. Suggestions for further research should be discussed. Finally, the investigator might allude to the social and medical impact of the study findings. How many lives can be saved? How many working days will be gained? Can symptoms be alleviated? Economic implications are important. Any benefit has to be weighed against the cost and feasibility of use in routine medical practice rather than in the special setting of a clinical trial.

REFERENCES

1. Glantz, S.A. Biostatistics: how to detect, correct and prevent errors in the medical literature. *Circulation* 61:1–7, 1980.
2. Relman, A.S. What a good medical journal does. *The New York Times.* Section IV; p. 22, March 19, 1978.
3. Sackett, D.L. Design, measurement and analysis in clinical trials. Edited by J. Hirsh, J.F. Cade, A.S. Gallus, and E. Schoenbaum. In *Platelets, Drugs and Thrombosis.* Basel: Karger, 1975.
4. Gifford, R.H., and Feinstein, A.R. A critique of methodology in studies of anticoagulant therapy for acute myocardial infarction. *N Engl J Med.* 280:351–357, 1969.

5. Chalmers, T.C., Smith, H. Jr., Blackburn, B., et al. A method for assessing the quality of a randomized control trial. *Controlled Clin Trials* 2:31–49, 1981.

6. Mosteller, F., Gilbert, J.P., and McPeek, B. Reporting standards and research strategies for controlled trials: agenda for the editor. *Controlled Clin Trials* 1:37–58, 1980.

7. Karlowski, T.R., Chalmers, T.C., Frenkel, L.D. et al. Ascorbic acid for the common cold: a prophylactic and therapeutic trial. *JAMA* 231:1038–1042, 1975.

8. Report from the Committee of Principle Investigators. A co-operative trial in the primary prevention of ischaemic heart disease using clofibrate. *Br Heart J.* 40:1069–1118, 1978.

CHAPTER 19

Multicenter Trials

A multicenter trial is a collaborative effort which involves more than one independent center in the tasks of enrolling and following study subjects. Early contributions to the design of these trials were made by Hill.[1] Greenberg[2] provided a general discussion of methodology.

There has been a dramatic increase in the number of multicenter trials in the last two decades. According to the 1979 NIH Inventory of Clinical Trials, of the 986 trials being sponsored by the National Institutes of Health that year, about 15% were multicenter.[3] Of course, the size of these varied, depending on the requirements of the study. Multicenter studies are more difficult and more expensive to perform than single-center studies, and they bring perhaps less professional reward due to the requirement to share credit among many investigators. Nevertheless, they are carried out because of the need for them.[4] Levin and colleagues provide many examples of "the importance and the need for well-designed cooperative efforts to achieve clinical investigations of the highest quality."[5] The development, organization, and conduct of one multicenter trial, the Coronary Drug Project, has been fully discussed in a monograph.[6] This chapter will discuss the reasons why such studies are conducted and briefly review some steps in their planning, design and conduct.

FUNDAMENTAL POINT

Anyone responsible for organizing and conducting a multicenter study should have a full understanding of the complexity of the undertaking. Problems in conduct of the trial most often originate from inadequate and unclear communication between the participating investigators, all of whom must agree to follow a common study protocol.

REASONS FOR MULTICENTER TRIALS

1. The main rationale for multicenter trials is to recruit the necessary number of subjects within a reasonable time. Many clinical

trials have been — and still are — performed without a good estimate of the number of subjects likely to be required to test adequately the main hypothesis. Yet, if the primary response variable is an event which occurs relatively infrequently, or small group differences are to be detected, sample size requirements will be large (Chapter 7).

Studies requiring hundreds of subjects usually cannot be done at one center. For example, in the Aspirin Myocardial Infarction Study,[7] 30 centers enrolled the necessary 4200 patients with a history of a heart attack in one year and followed them for an additional three years. The largest of these centers enrolled slightly over 200 subjects. If an investigator were interested only in the experience of the subjects over the initial three years after enrollment, assuming no further benefit from intervention after that time, then the single largest center would have required 21 years to recruit subjects and 24 years to complete the study. Even if the investigator were interested simply in an equivalent number of person-years of intervention, regardless of the number of years a subject received the intervention, this one center would have taken approximately 12 years to complete the study. This assumes uniform annual rates of enrollment over nine of these 12 years, uniform annual mortality over the 12 years, and follow-up of all subjects to a common termination date.

Even a 12-year study may be impractical and may develop major problems. New advances in therapy and methodology during the years may make the study obsolete. Mortality from causes other than the one of interest may become more important in the later years of the study, diluting any effect of the intervention. It may not be reasonable to expect an intervention to continue to provide the same relative benefit over the course of many years. In addition, subjects and investigators are likely to lose interest in the trial and elect not to participate further. There is also a good chance that they may move from the area. Finally, answers from the trial which might benefit other people will be delayed. For these reasons, most investigators prefer to engage in a study of shorter duration.

2. A multicenter study may assure a more representative sample of the study or target population. Geography, race, socioeconomic status and life style are among the factors which can sometimes be more representative of the general population if many centers enroll subjects than if only one does so. These factors may be important in the ability to generalize the findings of the trial. Severity and sequelae of hypertension, for example, are seemingly race related. A study of hypertensive patients from either a totally black or totally white community could yield findings not necessarily applicable to a more diverse population. Similarly, a study of pulmonary disease in an air-polluted industrial center might not give the same results as a study in a rural area.

3. A multicenter study enables investigators with similar interests

and skills to work together on a common problem. Science and medicine, like many other disciplines, are competitive. Nevertheless, investigators may find that there are times when their own interests, as well as those of science, require them to cooperate. Thus, many scientists collaborate in order to solve particularly vexing clinical problems and to advance knowledge in areas of common interest. A multicenter trial also gives capable, clinically oriented persons from a variety of institutions, who might otherwise not become involved in research activities, an opportunity to contribute to science.

CONDUCT OF MULTICENTER TRIALS

One of the earlier multicenter clinical trials was the Coronary Drug Project.[6] This study provided an initial model for many of the techniques currently employed. Some techniques have been refined in subsequent trials. As in all active disciplines, concepts are frequently changing. However, at this time, the following outline or series of steps is one possible way to approach the planning and conduct of a multicenter trial. It consists of a distillation of experience from a number of these studies.

First, a group should be established to be responsible for organizing and overseeing the various phases of the study (planning, subject recruitment, subject follow-up, phaseout, data analysis, paper writing) and its various centers and committees. This group often consists of people in government agencies, private research organizations, educational institutions, or private industry, with input from appropriate consultants. Use of consultants who are expert in the field of study, in biostatistics, and in the management of multicenter clinical trials is encouraged. The organizing group needs to have authority in order for it to operate effectively and for the study to function efficiently.

Second, to determine the feasibility of a study, the organizing group should make a thorough search of the literature and review of other information. Sample size requirements should be calculated. Reasonable estimates must be made regarding control group event rate, anticipated effect of intervention and subject compliance to therapy. The organizing group also has to evaluate key issues such as total cost, subject availability, presence of competent cooperating investigators and timeliness of the study. After assessing these, is the trial worth pursuing? Are there sufficient preliminary indications that the intervention under investigation indeed might work? On the other hand, is there so much suggestive (though inconclusive) evidence in favor of the new intervention that it might be difficult to allocate subjects to a control group? Since planning for the study may take a year or more, feasibility needs constantly to be re-evaluated, even up to the time of the actual start of subject recruitment. New or impending evidence may at any time cause cancellation, postponement or redesign of the trial. In some instances, a pilot, or feasibility study is useful in answering specific questions important for the design and conduct of a full-scale trial.

Third, multicenter studies require not only clinical centers to recruit subjects, but also centers to perform specialized activities (eg, read electrocardiograms, perform key laboratory tests, read angiograms or x-rays, read pathology slides, distribute study drugs), and a coordinating center to help design and manage the trial and to collect and analyze data from all other centers. While the specialized centers may perform multiple services, it is usually not advisable to have any of the clinical centers that are involved perform these services. If a specialized center and a clinical center are in the same institution, each should have a separate staff. Otherwise, unblinding and (therefore, bias) could result. Even if unblinding or bias is avoided, there might be criticism that this could have occurred and thus raise unnecessary questions about the entire clinical trial.

As reported by Croke,[8] a major consideration when selecting clinical center investigators is availability of appropriate subjects. The trial has to go where the subjects are, not the other way around. Clearly, experience in clinical trials and scientific expertise are desirable features for investigators, but they are not crucial to overall success. Well-known scientists who add stature to a study are not always successful in collaborative ventures. The chief reason for this is their inability to devote sufficient time to the trial.

The selection of the coordinating center is of utmost importance. In addition to helping design the trial, this center is responsible for implementing the randomization scheme, for carrying out day-to-day trial activities, and for collecting, monitoring, editing, and analyzing data. The coordinating center needs to be in constant communication with all other centers. Its staff has to have expertise in areas such as biostatistics, computer technology, epidemiology, medicine and management in order to respond expeditiously to daily problems which arise in a trial. These might range from simple questions, such as how to answer a particular item on a questionnaire, to requests for special data analyses that may require modifications of established statistical methods. In all, the staff at the coordinating center has to be experienced, capable, responsive and dedicated in order to handle its workload in a timely fashion. A trial can succeed despite inadequate performance of one or two clinical centers, but a poorly performing coordinating center can materially affect the success of a multicenter trial. A key element in any coordinating center is not only the presence of integrity, but the appearance of integrity. Any suspicion of conflict of interest can damage the trial. It is for this reason that pharmaceutical firms who support trials sometimes use outside biostatistical staffs as coordinating centers. In all cases, the personnel in a coordinating center should be seen to have no overriding interest in the outcome of a trial.

It is also obvious that the centers selected to perform specialized activities need expertise in their particular fields. Equally important is the capacity to handle the usually very large workloads in a multicenter trial with research-level quality. Even with careful selection of these center, backlogs of work are a frequent source of frustration during the course of a trial.

Fourth, it is preferable for the organizing group to provide prospective investigators with a fairly detailed outline of the key elements of the study design as early as possible. This results in more efficient initiation of the trial and allows each investigator to better plan his staffing and cost requirements. Rather than presenting a final protocol to the investigators, it is recommended that they be given time to discuss and, if necessary, modify the trial design. This allows them to contribute their own ideas, to have an opportunity to participate in the design of the trial and to become familiar with all aspects of the study. The investigators need a protocol which is acceptable to them and their colleagues at their local institution. Usually several planning sessions prior to the start of subject recruitment are needed for this process.

If there are a great many investigators and a number of difficult protocol decisions, it is useful during the planning stage to have different investigators address these issues. Working groups can focus on individual problems and prepare reports for the total body of investigators. Of course, if the initial outline has been well thought out and developed, few major design modifications will be necessary. Any design change needs to be carefully examined to ensure that the basic objectives and feasibility of the study are not threatened. This applies particularly to modifications of subject eligibility criteria. Investigators are understandably concerned about their ability to enroll a sufficient number of subjects. In an effort to make recruitment easier, they may favor less stringent eligibility criteria. Any such decisions need to be examined to ensure that they do not have an adverse impact on the objectives of the trial and on sample size estimates. The benefit of easier recruitment may be outweighed by the need for a larger sample size. Planning meetings serve to make all investigators aware of the wide diversity of opinions. Inevitably, compromises consistent with good science must be reached on difficult issues, and some investigators may not be completely satisfied with all aspects of a trial. However, all are usually able to support the final design. *All investigators in a cooperative trial must agree to follow the common study protocol.*

Fifth, an organizational structure for the trial should be established with clear areas of responsibility and lines of authority. Many have been developed. [9-11] The one below is based upon the Beta-Blocker Heart Attack Trial, and is similar in structure to the others.

Policy and Data Monitoring Board This scientific body, which should be independent of the investigators and any sponsor of the trial, is charged with periodically monitoring baseline, toxicity, and response variable data and evaluating center performance. [12] It reports to either the organizing group or the study sponsor (which may be one and the same). In different studies, the functions of the Board have been performed by two separate committees, by one committee or by one committee with a subcommittee. The use of two, one to review data and another to advise on policy matters, has the theoretical advantage of providing an additional review. However, if the policy advisory group also has direct responsibility for data review, it can discharge its duties more effectively. That is because it has a more intimate knowledge of study data and can directly request from the coordinating center any additional data analyses which might assist it during its deliberations. Therefore, the trial should have either a single committee or a subcommittee of the main group which is responsible for data review and must report to the committee as a whole. The coordinating center should present tabulated and graphic data and appropriate analyses to the Policy and Data Monitoring Board for review. The Board has the responsibility to recommend to the organizers or to the sponsor of the trial early termination in case of unanticipated toxicity or greater-than-expected benefit. (See Chapter 15.) Members of this Board should be knowledgeable in the field under study, in clinical trials methodology, and in biostatistics. Some people also suggest adding a subject advocate to the group. The qualifications of this person will vary, depending on the nature of the trial. The ethical responsibilities of the Policy and Data Monitoring Board to the subjects, as well as to the integrity of the study, should be clearly established. These responsibilities for subject safety are particularly important in double-blind studies, since the individual investigators are unaware of the group assignments.

Steering Committee In large studies, this committee provides scientific direction for the study at the operational level. Its membership is made up of a subset of investigators participating in the trial. Depending on the length of the study, some key investigators may be permanent members of the Steering Committee to provide continuity. Others may be chosen or elected for shorter terms. Subcommittees are often established to consider on a study-wide level specific issues such as compliance, quality control, classification of response variables, and publication policies and then report to the Steering Committee.

It is also important to authorize a small subgroup to make executive decisions between Steering Committee meetings. Most "housekeeping" tasks can be more easily accomplished in this manner. A large committee, for example, is unable to monitor a trial on a daily basis, write memoranda, or prepare scientific papers. Since committee meetings can rarely be called at short notice, issues requiring rapid decisions must be addressed by an executive group. It is important, however, that major questions be discussed with the investigators.

Assembly of Investigators   This committee represents all of the centers participating in the trial. In small studies, this Assembly may equate to the Steering Committee. In large studies, the Steering Committee would become too large to perform its duties effectively if it were made up of all investigators. In this model, the Principal Investigator for each center is a voting member of the Assembly. The purposes of Assembly meetings, which may be attended by other study personnel, are to allow for votes on major issues, to keep all investigators acquainted with the progress of the trial, and to provide an opportunity for staff training and education. Given the complexity of many trials, this last purpose is often the most important.

The committee structure for the multicenter Beta-Blocker Heart Attack Trial[9] was a fairly typical example. Two subcommittees (Mortality Classification and Nonfatal Events) dealt with evaluation of response variables. Such central evaluation, with the subject's identity and intervention group assignment blinded, helped to assure unbiased classification of reported events and to eliminate problems of variable interpretation of event definition. Other subcommittees such as Compliance and Quality Control were responsible for assuring high quality data by monitoring performance and initiating corrective action if needed. Usually, there are subcommittees which develop and review study publications and presentations (Editorial Subcommittee, Natural History Subcommittee, and Bibliography Subcommittee). Finally, there are subcommittees with specialized functions depending on the nature of the trial. In the Beta-Blocker Heart Attack Trial, a subcommittee addressed issues concerning conduct and analysis of ambulatory electrocardiographic recordings. These bodies were purposely called subcommittees to emphasize that they are responsible to the Steering Committee.

In some other trials, the committee structure has become too complex, leading to inefficiencies. If committees, subcommittees and task forces multiply, the process of handling routine problems becomes difficult. Studies which involve multiple disciplines especially need a carefully thought out organizational structure. Investigators from different fields tend to look at issues from various perspectives. Although this can be beneficial, under some circumstances it can obstruct the orderly conduct of a trial. Investigators may seek to increase their own areas of responsibility and, in the process, change the scope of the study. What starts out as a moderately complex trial can end up being an almost unmanageable undertaking.

Sixth, despite special problems, multicenter trials should try to maintain standards of quality as high as those attainable in carefully conducted single center trials. Therefore, strong emphasis should be placed on training and standardization. It is obviously extremely important that staff at all centers understand the protocol definitions, and how to complete forms and perform tests. Differences in performance from center to center, as well as between individuals in a single center, are unavoidable. They can, however, be minimized by proper training, certification procedures, retesting, and when necessary, retraining of staff. These efforts need to be implemented before a trial gets underway. (See Chapter 10 for a discussion of quality control.) Not until the Steering Committee is satisfied that staff are capable of performing necessary procedures should a clinical center be allowed to begin enrolling subjects. Meetings of the Assembly of Investigators are essential to the successful conduct of the trial because they provide opportunities to discuss common problems and review proper ways to collect data and complete study forms. In one trial, very elaborate and extensive standardization, certification, and periodic retesting was undertaken to ensure that methodology and procedures for laboratory determinations were identical at the various centers.[13] Such standardization was rather costly but essential to the completion of the goals of that study.

Certain functions in a multicenter trial are best carried out by properly selected special centers. The advantages of centrally coding electrocardiograms, reading x-rays, evaluating pathology specimens, or performing laboratory tests include unbiased assessment, reduced variability and high quality performance. The disadvantages of centralized determinations include the cost and time required for shipping, as well as the risk of losing study material.

Seventh, there needs to be close monitoring of the performance of all centers. Subject recruitment, quality of data collection and processing, quality of laboratory procedures and compliance of subject to protocol should be evaluated frequently. Exactly how often is determined by the time span for which investigators are willing to allow errors or nonperformance to go undetected. Of course, cost and personnel considerations may dictate lesser frequency than desired. Tables 19-1 through 19-4 are typical of the kinds of information that investigators have used to compare clinical center performance. Table 19-1 compares the average time it takes for each clinical center to complete and send in forms to the coordinating center. Center B in the present six months stands out as doing poorly and it has become worse when compared with previous performance. Table 19-2 shows that center B has a large number of unsubmitted forms. As seen in Table 19-3, clinical centers A and B are submitting many forms with errors. Table 19-4 illustrates that centers have problems in supplying laboratory data. It is useful to identify those centers which are performing below average. Often, specific problems can be identified and corrected. For example, in one study, evidence of left ventricular hypertrophy was identified much more often in the electrocardiograms from one center than from the other centers. Only after looking into the reasons for this was it discovered that the internal standard on that clinical center's electrocardiograph machine was incorrectly calibrated.

In all clinical trials, recruitment of subjects is difficult. In a cooperative clinical trial, however, there is an opportunity for some clinical centers to compensate for the inadequate performance of other centers. The clinical centers should understand that, while friendly competition keeps everybody working, the real goal is overall success, and

Table 19-1
Average Processing Time for Follow-up Visit (FV) Forms*

Clinic	Previous Six Months		Present Six Months	
	No. of FV forms received	*Days from visit to receipt of forms*	*No. of FV forms received*	*Days from visit to receipt of forms*
A	292	25.8	290	8.7
B	157	22.9	117	29.0
C	210	16.0	198	16.2
D	174	11.6	173	10.4
E	182	8.3	185	12.7
Total	1015	17.8	963	13.8

*Table used in Aspirin Myocardial Infarction Study: Coordinating Center, University of Maryland.

Table 19-2
Number of Follow-up Visit Forms Not Received at Coordinating Center More Than One Month Past the Visit Window, by Clinic*

Clinic	January 13, 1978	July 28, 1978
A	8	0
B	21	65
C	0	1
D	1	0
E	0	0
Total	30	66

*Table used in Aspirin Myocardial Infarction Study: Coordinating Center, University of Maryland.

294

what some centers cannot do, another perhaps can. Therefore, it is important to encourage the good centers to recruit as many subjects as possible.

Eighth, publication, presentation and authorship policies should be agreed upon in advance. Authorship becomes a critical issue when there are multiple investigators, many of whose academic careers depend on publications. Unfortunately, there is no completely satisfactory way to recognize the contribution of each investigator. A common compromise is to put the study name immediately under the paper title and to acknowledge the writers of the paper, either in a footnote or under the title, next to the study name. All key investigators are then listed at the end of the paper. The policy may also vary according to the type of paper (main or subsidiary).

In one four-center trial, the investigators at one of the centers reported their own findings before the total group had an opportunity to do so.[14,15] Such an action is not compatible with a collaborative effort. It undermines the goal of a multicenter trial of having enough subjects to answer a question and, perhaps more important, the trust among investigators. Those unwilling to abide by the rule for common authorship should not participate in collaborative studies.

Advance planning of authorship policy may eliminate subsequent misunderstandings. However, fair recognition of junior staff will always be difficult.[16] Study leadership often gets credit and recognition for work done largely by people whose contributions may remain unknown to the scientific community. One way to alleviate this problem is to appoint as many capable junior staff as possible to subcommittees. Such staff should also be encouraged to develop studies ancillary to the main trial. This will enable them to claim authorship for their own work while using

Table 19-3
Percent of Follow-up Visit Forms With One or More Errors, by Clinic and Month of Follow-up*

Clinic	Feb.	Mar.	Apr.	May	June	July	Total	Total Previous 6 Months	Errors Per Form†	No. of Forms Processed
A	30.0	29.6	29.6	29.7	35.1	29.3	30.3	33.2	6.11	290
B	25.0	14.3	20.8	28.0	—	28.3	24.8	24.2	6.66	117
C	0.0	14.1	3.4	8.1	27.3	13.8	12.1	16.2	5.21	198
D	22.2	16.7	6.3	20.9	9.7	26.3	16.2	17.8	6.21	173
E	4.8	10.3	19.6	13.6	21.2	20.8	15.7	18.7	4.38	185
Total	17.0	18.5	17.0	20.4	20.9	23.8	20.6	23.1	5.68	963

*Table used in Aspirin Myocardial Infarction Study: Coordinating Center, University of Maryland.
†Errors per form are calculated by dividing the number of errors by the number of forms failing edit.

Table 19-4
Number and Percent of Missing or Non-Valid Follow-up Laboratory Chemistries and Annual ECGs, by Clinic, All Follow-up Visits Combined*

Clinic	Platelet Aggregation		Annual Serum Chemistries		Urine Salicylate		Annual Electrocardiogram	
	No.	%	No.	%	No.	%	No.	%
A	57	4.59	2	0.58	4	0.28	0	0.00
B	137	22.64	6	3.55	4	0.55	0	0.00
C	13	1.49	0	0.00	4	0.39	0	0.00
D	55	7.54	5	2.27	4	0.47	0	0.00
E	99	12.36	20	9.22	3	0.32	0	0.00
Total	361	8.50	33	2.81	19	0.38	0	0.00

*Table used in Aspirin Myocardial Infarction Study: Coordinating Center, University of Maryland.

the basic structure of the trial to get access to subjects and supporting data. Such ancillary studies may be performed on only a subgroup of subjects and may not necessarily be related to the trial as a whole. Care must be taken to ensure that they do not interfere with the main effort, either through unblinding, by harming the subjects or by causing the subjects to leave the trial.

GENERAL COMMENTS

Even if investigators think they have identified all potential difficulties and have taken care to prevent them, new problems will always arise. This is particularly true of multicenter trials because of their size, complexity and large number of investigators with diverse backgrounds and interests. To forestall and minimize problems, the need for adequate study-wide communication must be stressed. If communication between the various components of the study lapses, or is vague, the trial can rapidly deteriorate. It is the responsibility of the coordinating center to keep in frequent contact (by telephone, letter, and visit) with all the other centers. This contact needs to be initiated by both the coordinating center and the other centers. The study leaders also need to maintain contact with the various centers and committees, closely monitoring the conduct of the trial.

Cost is always a concern in multicenter trials. These studies are generally expensive due to their complexity, size and elaborate committee structure. Expense can be minimized by asking only pertinent questions on forms, by reducing the number of laboratory tests and by performing only necessary quality monitoring; in essence, by simplifying data collection.[17] Accomplishing these economies demands constant attention, particularly during the planning phase. The investigators in a multicenter trial traditionally represent diverse interests. Given the opportunity, most of them would pursue these interests. Few would like to miss an opportunity to add to the main trial questions or examinations of particular importance to them. These additions are often important scientifically. However, it is easy for a trial to become overbuilt and get out of control. It is usually a good policy to restrict additions to the basic study protocol. Special caution should be taken when the argument for inclusion is, "it would be interesting to know." The purpose of every procedure and item on the study forms should be clearly defined in advance and a test hypothesis formulated, if possible. Certain questions can be answered by using less than the total number of subjects. Other questions require the whole group. Still others cannot be answered, even if all subjects are included. Therefore, in each instance, thought should be given to sample size and power calculations.

REFERENCES

1. Fleiss, J.L. Multicentre clinical trials: Bradford Hill's contributions and some subsequent developments. *Stat Med*. 1:353–359, 1982.

2. Greenberg, B.G. Conduct of cooperative field and clinical trials. *Am Stat*. 13:13–28, June 1959.

3. NIH Inventory of Clinical Trials: Fiscal Year 1979. Volume I. National Institutes of Health, Division of Research Grants, Research Analysis and Evaluation Branch, Bethesda, MD.

4. Klimt, C.R. Principles of multi-center clinical studies. Edited by J.P. Boissel and C.R. Klimt. In *Multi-center Controlled Trials. Principles and Problems*. Paris: INSERM, 1979.

5. Levin, W.C., Fink, D.J., Porter, S. et al. Cooperative clinical investigation: a modality of medical science. *JAMA* 227:1295–1296, 1974.

6. Coronary Drug Project Research Group. P.L. Canner, (Ed). The Coronary Drug Project: methods and lessons of a multicenter clinical trial. *Controlled Clin Trials* 4:273–541, 1983.

7. Aspirin Myocardial Infarction Study Research Group. A randomized, controlled trial of aspirin in persons recovered from myocardial infarction. *JAMA* 243:661–669, 1980.

8. Croke, G. Recruitment for the National Cooperative Gallstone Study. *Clin Pharmacol Ther*. 25:691–694, 1979.

9. Byington, R.P., for the Beta-Blocker Heart Attack Trial Research Group. Beta-Blocker Heart Attack Trial: design, methods, and baseline results. *Controlled Clin Trials* 5:382–437, 1984.

10. Meinert, C.L. Organization of multicenter clinical trials. *Controlled Clin Trials* 1:305–312, 1981.

11. Lachin, J.M., Marks, J.W., Schoenfield, L.J., et al. Design and

methodological considerations in the National Cooperative Gallstone Study: a multicenter clinical trial. *Controlled Clin Trials* 2:177–229, 1981.

12. Friedman, L. and DeMets, D. The data monitoring committee: how it operates and why. *IRB* 3:6–8, 1981.

13. Lipid Research Clinics Program Manual of Laboratory Operations: Volume I. Lipid and Lipoprotein Analysis. Washington, D.C.: DHEW Publication No. (NIH)75-628, 1975.

14. Winston, D.J., Ho, W.G., and Gale, R.P. Prophylactic granulocyte transfusions during chemotherapy of acute nonlymphocytic leukemia. *Ann Intern Med*. 94:616–622, 1981.

15. Strauss, R.G., Connett, J.E., Gale, R.P., et al. A controlled trial of prophylactic granulocyte transfusions during initial induction chemotherapy for acute myelogenous leukemia. *N Engl J Med*. 305:597–603, 1981.

16. Remington, R.D. Problems of university-based scientists associated with clinical trials. *Clin Pharmacol Ther*. 25:662–665, 1979.

17. Yusuf, S., Collins, R. and Peto, R. Why do we need some large, simple, randomized trials? *Stat Med*. 3:409–420, 1984.

INDEX